FROM COMPUTING TO COMPUTATIONAL THINKING

FROM COMPUTING TO COMPUTATIONAL THINKING

Paul S. Wang
Kent State University
Ohio, USA

CRC Press is an imprint of the
Taylor & Francis Group, an **informa** business
A CHAPMAN & HALL BOOK

CRC Press
Taylor & Francis Group
6000 Broken Sound Parkway NW, Suite 300
Boca Raton, FL 33487-2742

Printed at CPI UK on sustainably sourced paper
Version Date: 20150624

International Standard Book Number-13: 978-1-4822-1765-0 (Paperback)

Visit the Taylor & Francis Web site at
http://www.taylorandfrancis.com

and the CRC Press Web site at
http://www.crcpress.com

Contents

Preface

It has been widely recognized that concepts, techniques, and analytical abilities from the field of computing can be powerful mental tools in general for solving problems, performing tasks, planning, working with others, anticipating problems, troubleshooting, and more. We refer to this mental tool set as *computational thinking* (CT).

This textbook will help readers acquire computational thinking through an understanding of modern computer technologies. Neither programming background nor learning how to program is required. Students just need to bring their curiosity and an open mind to class.

Reading this book can be an excellent way to prepare someone to pursue a rewarding career in computing or information technology. The materials are as much about computing as about sharpening the mind.

Topics and Presentation

The book has an end-user viewpoint. Topics are presented in an interesting and thought-provoking way, keeping the reader engaged and motivated to continue.

Unconventional chapter titles, CT callout boxes, relation to daily living, and connection to well-known events combine to encourage computational thinking and instill agile mental skills. In addition, we introduce a new verb in English, *computize*. To computize is to apply CT. With a little bit of help, everyone can do it.

The CT callout boxes highlight nuggets of computational thinking wisdom worth revisiting from time to time. They can be found easily in the Table of Contents and in the Index.

The user is guided through a well-selected set of topics covering the type of material appropriate for a one-semester course at the college freshman level for students from all different majors. Advanced programs in high schools, and the public in general, may find this book useful and rewarding as well.

Computing and CT

Understanding computing and acquiring CT are two sides of the same coin. By learning about hardware, software, networking, the operating system, security measures, the Web, digital data, apps, and programming paradigms, we gain valuable knowledge to better take advantage of information technologies.

At the same time, concepts and methods from computing form elements of CT that are applicable outside of computing. CT can make us wiser and more effective in countless ways. CT can help us avoid accidents and mishaps. It can even be life saving.

Chapter-end exercises reinforce topics in each chapter and challenge students to apply CT (to computize) in various situations. Group discussions are encouraged as well.

The CT Website

Throughout the book, concepts, techniques, and technologies are explained with many interesting examples. Hands-on demos for experimentation are online at the book's companion website `http://computize.org`

The site is mobile-enabled and works on both regular and mobile devices. In the text, we refer to it as *the CT website*. The live demos are cross-referenced to in-text descriptions with a notation such as **Ex:UpCounter** that also appears in the book's index.

The CT website offers additional resources, allows you and others to share insights on CT, and provides information updates.

Acknowledgments

I thank my dear wife, Jennifer Wang (葛孝薇), who encouraged me to embark on this project, read all drafts, and provided great feedback, sometimes with specific ideas and wording changes. I am very grateful to her.

In fact, she was the one who asked me to listen to a broadcast of the *Kojo Nnamdi Show* (WAMU/NPR) on November 18, 2008. The show started with an interview of Dr. Jeannette Wing on the topic

"*Thinking like a computer scientist.*"

Jennifer was so excited and told me over the phone, "You need to listen to this right now, it is what you always talked about."

Encouraged by Dr. Wing's advocacy, I soon contacted her at the National Science Foundation and invited her to visit our Computer Science Department at Kent State University to give a talk on CT. The face-to-face interactions with Dr. Wing further convinced me to make a contribution in this important direction. The influence of NPR and Dr. Wing on me cannot be overstated.

Thanks must also go to my children Laura, Deborah, and David, who took an active interest in this book; especially David who read first chapters of an early draft and made comments that influenced the choice of the book title.

I also thank Randi Cohen, editor at CRC Press, who supported this project with energy and enthusiasm from the very beginning. She also helped to choose the book title.

Deep thanks also go to production coordinaters, Kathryn Everett and Amber Donley as well as project editor Robin Lloyd-Starkes, and others at CRC Press for their professionalism and dedication.

During the planning and writing of this book, several reviews have been conducted. Much appreciated are the input and suggestions from the reviewers:

- Iyad A. Ajwa, Ashland University, Ohio, USA
- Lian Li (李廉教授) and his team, HeFei University of Technology, Anhui, China
- Alex Melton, Benjamin Logan High School, Ohio, USA
- Anonymous reviewers

Paul S. Wang
王 士 弘
Kent, Ohio
pwang@cs.kent.edu

Introduction

Digital computers brought us the information revolution. Citizens in the information age must deal with computers, smartphones, and the Internet. In addition, they also need to gain *computational thinking.*

Computational thinking (CT) is the mental skill to apply fundamental concepts and reasoning, derived from modern digital computers and computer science, in all areas, including day-to-day activities. CT is thinking inspired by an understanding of computers and information technologies, and the advantages, limitations, and problems they bring. CT also encourages us to keep asking questions such as "*What if we automate this?*" "*What instructions and precautions would we need if we were asking young children to do this?*" "*How efficient is this?*" and "*What can go wrong with this?*"

CT can expand your mind, help you solve problems, increase efficiency, avoid mistakes, and anticipate pitfalls, as well as interact and communicate better with others, people or machines. CT can make you more successful and even save lives!

It is not necessary to become a computer scientist or engineer for you to acquire CT. From a user point of view, we present a well-organized sequence of topics to introduce computers and computing in simple and easy ways, assuming little prior knowledge of computer science or programming. While presenting the hardware, software, data representation, algorithm, systems, security, networking, the Web, and other aspects of computing, we will highlight widely applicable concepts and mental skills in *CT call-out boxes* and explain how/where they can be applied in real life.

Background

Back in March 2006, Dr. Jeannette M. Wing published an article on computational thinking in *The Communications of ACM* and boldly advocated it as a skill for everyone:

> "Computational thinking builds on the power and limits of computing processes, whether they are executed by a human or by a machine. Computational methods and models give us the courage to solve problems and design systems that no one of us would be capable of tackling alone. ... Computational thinking is a fundamental skill for everyone, not just for computer scientists. To reading,

writing, and arithmetic, we should add computational thinking to every child's analytical ability. Just as the printing press facilitated the spread of the three Rs, what is appropriately incestuous about this vision is that computing and computers facilitate the spread of computational thinking."

Within the academic research community, there have been significant discussions on computational thinking, what it encompasses, and its role inside the education system.

In educational circles, there is an increasing realization of the potential importance of learning to think computationally. According to a recent report on computational thinking by the National Research Council of The National Academies (NRC):

> "... Computational thinking is a fundamental analytical skill that everyone, not just computer scientists, can use to help solve problems, design systems, and understand human behavior. ... Computational thinking is likely to benefit not only other scientists but also everyone else. ..."

The ACM/IEEE-CS Joint Task Force on Curriculum recently (2013) stated

> "Computational Thinking—While there has been a great deal of discussion in regard to computational thinking, its direct impact on curriculum is still unclear. While we believe there is no 'right answer' here, CS 2013 seeks to gain more clarity regarding models by which CS curricula can promote computational thinking for broader audiences."

Discovering the Secrets of CT

Computers are dumb. They deal only with *bits*. Each bit represents either a zero or a one. They blindly follow program instructions and operate on data, both being represented by sequences of `0`s and `1`s. Yet, they are universal machines that can perform any tasks when given instructions. The ways they are programmed, controlled, and made to work are fascinating to learn by themselves. But such understanding has more to give us, namely, computational thinking.

Important aspects of CT include

- Simplification through abstraction—Abstraction is a technique to reduce complexity by ignoring unimportant details and focusing on what matters. For example, a driver views a car in terms of how to drive it and ignores how it works or is built. A user cares only about which mouse button to click and keys to press and generally overlooks how computers work internally.

- Power of automation—Arranging matters so they become routine and easy to automate. Working out a systematic procedure, an algorithm, for carrying out recurring tasks can significantly increase efficiency and productivity.

- Iteration and recursion—Ingeniously reapplying the same successful techniques and repeatedly executing the same set of steps to solve problems.

- An eye and a mind for details—Changing a 0 to a 1, or an upper-case `O`, can mess up the whole program. You need eyes of an eagle, mind of a detective, and a careful and meticulous approach. Overlooking anything can and will lead to failure.

- Precision in communication—Try telling the computer to do what you mean and not what you say `;-)`. You need to spell it out precisely and completely. Don't spare any details. Vagueness is not tolerated. And contexts must be made explicit.

- Logical deductions—"Cold logic" rules. Causes will result in consequences, whether you like it or not. There is no room for wishful or emotional thinking.

- Breaking out of the box—A computer program executes code to achieve any task. Unlike humans, especially experts, it does not bring experience or expertise to bear. Coding a solution forces us to think at a dumb computer's level (as if talking to a one-year-old) and get down to basics. This way, we will naturally need to think outside any "boxes."

- Anticipating problems—Automation relies on preset conditions. All possible exceptions must be met with prearranged contingencies. Ever said "I'll take care of that later"? Because there is a chance you might forget, according to CT, you should have a contingency plan ready in case you do forget. Otherwise, you have set a trap for yourself.

These are just some of the main ideas. CT offers you many more concepts and ways to think that can be just as, if not more, important. With increasing understanding of computing, one begins a process of gaining CT insights from many angles and viewpoints. Such CT takeaways can vary from person to person, in terms of what they are and their significance.

Here is a chicken and egg question. Which comes first, computing or computational thinking? Surely, ideas and techniques, from other disciplines as well as the long history of human civilization, have contributed to the development, breakthroughs, and refinements in computing. Yet, computer science has also generated many unique concepts, techniques, and problem-solving ideas. Computing has given rise to a digital ecosystem, called *cyberspace*, that includes us all.

Understanding the digital computer and computation is beneficial in itself. Plus, it gives us a very efficient way to discover/rediscover a set of powerful ideas, collectively known as CT, that can be applied widely. As we gain more understanding of computing and its various aspects, we will raise CT ideas along the way. Repeat visits of CT concepts from different aspects and contexts of computing provide different viewpoints and help instill the concepts, their usefulness and generality, in our minds.

This textbook provides an interesting and thought-provoking way to gain general knowledge about modern computing. You'll be exposed to new notions and perspectives that not only enrich your thinking but also make you more successful. For example, without the notion of *germs*, people won't achieve proper hygiene practices or effectively prevent disease transmission. Similarly, without a general understanding of computing concepts, it is hard to become a full-fledged citizen of the digital age.

Taking ideas from one field and applying them in another is not new. In fact, many breakthroughs came from such interdisciplinary endeavors. For example, the new biomimicry science studies nature's models and then applies these designs, processes, and inspirations to solve our own problems.

Readers are encouraged to share their own views, insights, and inspirations on the CT website. How wonderful that we can use computing technology to join forces and help advance CT.

Computize

Definition: **computize**, verb. To apply computational thinking. To view, consider, analyze, design, plan, work, and solve problems from a computational perspective.

When considering, analyzing, designing, formulating, or devising a solution/answer to some specific problem, computizing becomes an important additional dimension of deliberation.

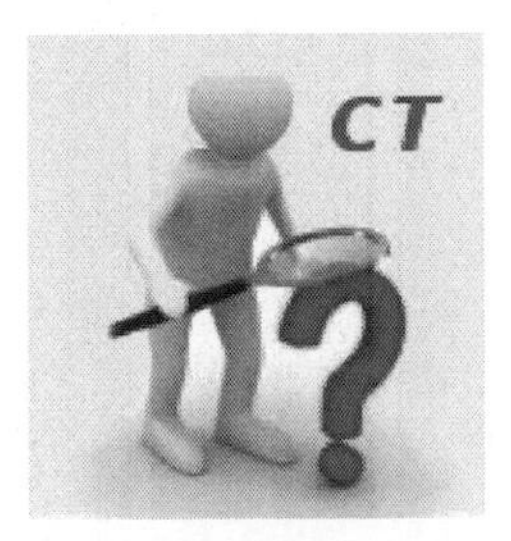

People say "hindsight is 20/20." But, since computer automation must deal with all possible applications in the future, we must ask "what if" questions and take into account all conceivable scenarios and eventualities. Let's look at a specific example. Hurricane Sandy was one of the deadliest and most destructive hurricanes in US history. With CT at multiple levels, dare we say that many of the disasters from Sandy might be substantially reduced.

- The New York City subway entrances and air vents are at street level. What if streets are flooded? What if flood water enters the subway?
- What if we need to fight fires in a flooded area? Do we have fire boats in addition to fire trucks? Do we have firefighters trained for boats?
- Most portable emergency power generators run on gasoline. What happens if gas runs out and gas stations are flooded?
- What if the drinking water supply stops? Can we provide emergency water from fire hydrants? In that case, can we use a mobile contraption that connects to a hydrant, purifies the water, and provides multiple faucets?
- What if emergency power generators are flooded? Should we waterproof generators in designated at-risk buildings?
- What if cell towers lose power? How hard is it to deploy airborne (drone?) cell relays in an emergency?
- What if we simulate storm damage with computer modeling and find out ahead of time what to prepare for?

So let's computize at multiple levels and do our best to get 20/20 hindsight beforehand. In the book, you'll find exercises to apply CT in various real or realistic situations.

Intended Use

By covering a well-selected set of topics, this text introduces computing to a general audience. Concepts and techniques in computing, in turn, inspire computational thinking. This approach can provide a deeper understanding of the computing concepts as well as reveal analytical abilities that apply well beyond computing. Students will learn new ways of thinking and problem solving.

Examples show how CT can impact day-to-day life and be applied to significant real-life problems. Readers will be exposed to useful thinking and problem solving skills of accomplished programmers without necessarily writing programs in any particular programming language.

The computer science community, with active encouragement from the professional societies (NRC, ACM, and IEEE Computer Society in particular), believes CT should be introduced into the academic curriculum, especially at the undergraduate level.

For example, the National Science Foundation (NSF)–funded project entitled "Piloting Pathways for Computational Thinking in a General Education Curriculum" (2008, Towson University) states

> "The vision is that 'Computational Thinking' becomes a standard general education category in the undergraduate curriculum at all academic institutions nationwide. To meet this goal, the main objective of this CPATH-CDP grant is to develop and assess a particular model for developing pathways of computational thinking throughout the general education (GenEd) curriculum at Towson University. Such pathways would be realized through the development of discipline-specific computational thinking courses by faculty members from various departments across all colleges on campus, supported by a common 'Everyday Computational Thinking' freshman-level course."

This textbook targets exactly such a common freshman-level course, perhaps offered by CS departments at the CS0 level. Students do not need a CS or programming background. Yet, it can easily lead students to computer science, computer engineering, or information technology. The book can also be used for computer literacy and other nonprogramming introductions to computing courses. Many parts of the book are also appropriate to introduce computing and computational thinking in the middle- or high-school level.

Because of its user viewpoint, the book can also be an interesting and rewarding read for the public at large.

Online Resources

The textbook has a companion website (the CT website) at

`computize.org`

where you will find many resources, including

- Interactive demos, cross-referenced inside the text, providing hands-on experience
- A place to share views, experiences, and insights on CT
- Information updates
- Classroom-ready lecture notes and other instructor resources

Chapter 1

Why Did the Chicken Cross the Road?

We are in the middle of an information revolution brought by the modern digital computer. It's nearly impossible for anyone to escape the reach of the resulting IT technologies. Even if you are not becoming a computer scientist or IT professional, a good overall understanding of how things work "under the hood" can be very advantageous, because we are all IT consumers.

The transforming power of the digital computer comes from it being a *universal machine*[1]. It is universal because it does not need rewiring or other physical modifications to perform different computations. Instead, the same hardware just needs to execute a different program. We'll look at the hardware and software aspects of the computer and its *finite state machine* model that help explain its universality.

By the way, the "Preface" and the "Introduction" provide important overviews, setting the tone for the entire textbook. Have you skipped them? If so, please consider going back before proceeding.

1.1 The Computer

The digital computer brought about the digital revolution, which started the information age. Increasingly, our daily activities depend on desktops, laptops, tablets, game consoles, and smartphones. They, in turn, rely on the Internet and the Web to provide instant access to information and online services worldwide. The global information infrastructure brings enormous benefits to the world economy, bridges geographical separations, enables instant interactions among people, and empowers individuals and entrepreneurs everywhere. It is fair to say that, for a modern citizen, computer literacy and computational thinking are as important as the ability to read, write, and do arithmetic.

Computers come in many forms. A desktop computer is a full-size, heavy-duty workhorse in the office and at home. Laptops are almost as powerful but can travel easily with you as well as provide integrated speakers, microphone, camera, and wi-fi access. Tablets are laptops without the keyboard or touch pad, making them even lighter and smaller. The touch screen opens up new and convenient ways to interact with the computer. A smartphone

[1]Here, the word machine refers to computing machines.

combines cellphone, tablet, and cell/wi-fi networking into a dream device for any information/communication junkie.

Furthermore, specialized computers run appliances, cars, ships, airplanes, spacecrafts, satellites, power grids, and robots. Their speed and precision enable us to build high-precision instruments, process enormous amounts of data quickly, and even intercept incoming missiles. Knowledge of how computers work and how to work with computers will help in almost any discipline or trade. And acquiring the mental skills for computational thinking will be beneficial in countless ways.

Computer Systems

Computer systems are easily among the most complex artifacts humans have ever built. The many aspects involved include: the central processing unit (CPU), graphics processing unit (GPU), cache memory, main random-access memory (RAM), secondary memory (hard disks), displays, keyboard, camera, microphone, speakers, printer, scanner, networking, operating systems, file systems, application programs, programming languages and compilers, user interfaces, protocols, file formats, user and process management, and so on. All these must interoperate and cooperate in precise and exact ways for the whole system to function properly. When testing or when things go wrong, sophisticated trouble-shooting and debugging systems and techniques are brought to bear. Bug fixes, security updates, and improvements, as well as new features, result in the constant release of new versions for these systems.

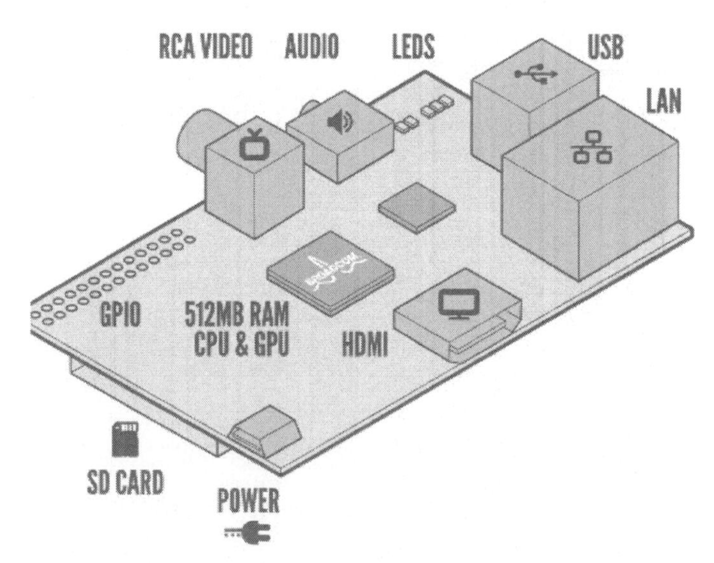

FIGURE 1.1 A Barebones Computer

Figure 1.1 shows a schematic for the Raspberry Pi, a credit-card sized computer that plugs into a TV and keyboard. It costs less than $40, runs the Linux operating system, and can be a great teaching/learning tool for computing.

The digital computer is quite simple at its core. The CPU executes a sequence of *instructions*, one after another. Each instruction has an *opcode*, representing the operation or action to be performed, and operands, values supplied for the operation. An operation may store a result in memory, perform I/O, designate which instruction to execute next, or otherwise change the *internal state* of the computer. Data and instructions are stored in memory in *binary codes* using zeros and ones. A computer program, however complicated or powerful, reduces to a sequence of instructions and data in binary (Section 10.5). What really distinguishes the modern digital computer from other machines is the fact that it is a *stored-program machine*. A computer instantly becomes a different machine depending on the program it executes. The fact that the same hardware can store and execute any possible programs on the fly makes the modern computer extremely powerful.

CPU instruction executions are synchronized by signals generated by an internal clock whose pulse frequency can range from several hundred MHz (10^6 Hz) to a few GHz (10^9 Hz)[2]. The faster the clock rate, the faster the CPU.

CPUs often have on-chip cache memory that is smaller, much faster to access, and more expensive than the main memory, usually RAM. Data and instructions from RAM are brought into cache to speed up repeated access. Caching is implemented in hardware, and cache management works in parallel with instruction execution to achieve faster speeds.

A CPU carries out instructions sequentially, one at a time. Modern computers can also increase their speed by having more than one CPU to carry out tasks *in parallel* or simultaneously.

1.2 Turing Machine

Alan Turing, considered by many as the father of computer science, was born on June 23, 1912, in England. His contributions are so important that Computer Science's Nobel Prize is named the *Turing Award.* At the tender age of 24, Turning published a paper introducing a computing model, now known as the *Turing machine.* A Turing machine is a theoretic device that models how a computer works (Figure 1.2).

A Turing machine can be described as follows.

- The machine has a finite number of internal states. It is a *finite-state machine* (FSM).

[2]Hz is a frequency unit. One Hz is 1 cycle per second.

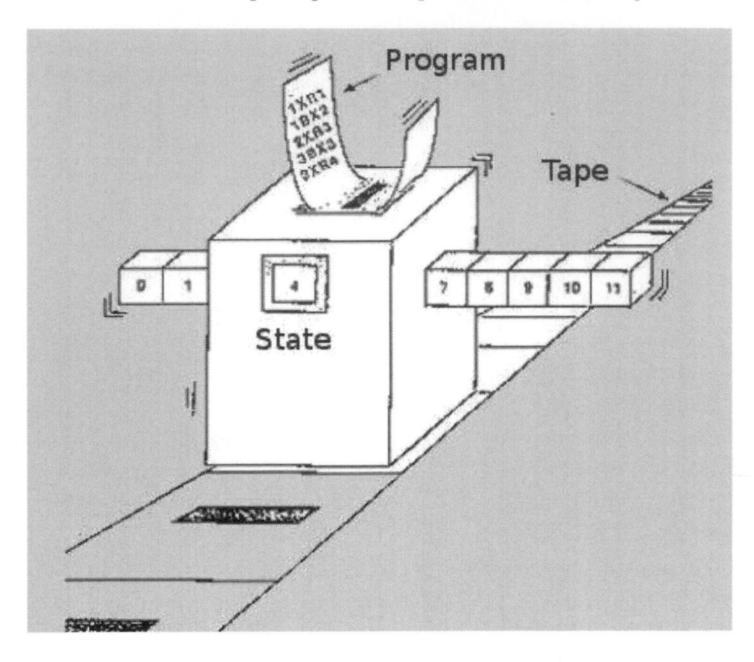

FIGURE 1.2 The Turing Machine

- Input to and output of the machine are given as symbols written on a tape that is not limited in length.
- The collection of symbols, the machine's *alphabet*, is a finite set.
- The symbol it is currently reading, together with its current state, determines the machine's action. An action may involve any and all of these: writing a new symbol on the tape, moving the read/write head forward or backward (or keeping it in place), entering a new state, and halting.
- The Turing machine's finite-length program specifies its different states, and exact actions for different input at each state, including state transitions.

Modern computers are shown to be equivalent to the Turning machine. The alphabet is the set {0, 1}, data and program stored in RAM define its states and actions, and it performs I/O.

Tasks that can be performed by a Turing machine are known as *computable* or *decidable*. Those that cannot be performed by a Turing machine are *incomputable* or *undecidable*. This simply means that a task is computable if it is possible to write a computer program for it.

The Turing machine pictured in Figure 1.2 starts working after loading a program. It has 12 states (0 through 11) and is currently in state 4.

The Turing machine is a prime example of a central idea in computer science–*abstraction.*

> **CT: ABSTRACT AWAY**
>
> *Strip away unimportant, unrelated, or irrelevant details from concrete instances, and focus on the essential core characteristics of the subject matter.*

Through abstraction, we often can make a complicated situation much clearer by peeling away unessential aspects that often can obscure the issue. The ability of abstraction allows us to distill the essence of an idea or concept and, therefore, gain a more profound understanding. For example, we can talk about adults, teenagers, children, liberals, conservatives, men, women, even numbers, odd numbers, and so on. But, not until we abstract to the concept of "set" and set membership, can we gain a clear concept and a language to discuss collections of objects in a general and rigorous manner.

As we make progress in this textbook, you will gain familiarity with abstraction, an important analytical skill that can be applied widely, not just for computing.

1.3 A Brief History of Computers

The need for devices to help with counting and arithmetic dates back to ancient times. A good example is the abacus, a simple and easy mechanical calculator still manufactured and in use today. With beads that slide on thin, smooth rods, the Chinese abacus was first described in a 190 CE book but perhaps in use much earlier. Figure 1.3 shows a typical Chinese abacus, where the number represented is 37925.

A more complicated mechanical calculator, the Charles Babbage *difference engine*, first built in 1823, uses gears and other precision parts to calculate values of polynomial functions to better make tables for mathematical functions. Babbage never totally completed his difference engine or the more advanced analytical engine. Based on his designs, models were later constructed for demonstration purposes. One, the Charles Babbage Difference Engine No. 2 (Figure 1.4), is on display at the the Computer History Museum in Mountain View, California, USA.

Like other complex devices, the modern digital computer came to be through a series of inventions and breakthroughs that, in turn, benefited from discoveries and contributions of many others earlier in history. The first programmable computer, the Z1, was invented by Konrad Zuse as early as 1936.

FIGURE 1.3 A Chinese Abacus (算盤)

FIGURE 1.4 Charles Babbage Difference Engine No. 2

In the period 1955–1975, computers used core memory as RAM. A core is a tiny magnetic ring that can be magnetized clockwise or counterclockwise to represent one bit of information. Major credits for developing the core memory system (Figure 1.5) in the late 1900s go to An Wang (王安), Way-Dong Woo (Harvard University), and Jay Forrester (MIT). Today, computer RAMs are semiconductor memory made of transistors in integrated circuits on chips. Working for Texas Instruments, Jack Kilby demonstrated the first working sample integrated circuit on September 12, 1958. Kilby won the 2000 Nobel Prize in Physics for his contribution to the invention of the integrated circuit. Integrated circuits are fabricated onto silicon wafers using a photographic printing process. Through the years, parts continuously became smaller and the scale of circuit integration greater. Today, an entire 64-bit microprocessor (CPU, GPU, ALU, and cache memory) with well over 1 billion transistors can be placed on a chip smaller than a thumbnail. Features on the chip are just a few nanometers (10^{-9} meter) in size.

Computer displays have evolved from teletype-/typewriter-like devices,

FIGURE 1.5 1024-bit Core Memory Plane

and character-oriented CRTs, to pixel-based, full-color, high-definition, touch-sensitive, LCDs and LEDs. Secondary storage moved from punched cards, paper tapes, magnetic drums (Figure 1.6), magnetic tapes, and floppy disks, to high-capacity hard disks, DVD/Blu-ray discs, flash drives, SD cards, and solid-state drives. The cost of secondary storage keeps going down.

FIGURE 1.6 A Magnetic Drum

1.4 Software

Every computer consists of two integral parts:

- Hardware—The physical devices that are difficult to alter or modify, including CPU, ALU (arithmetic logic unit), GPU, memory, disk drives, wired and wireless network adapters, and touch screen, as well as peripheral devices, such as monitor, keyboard, mouse, audio I/O, video camera, touch pad, printer, and so on.
- Software—Program code that can easily be loaded, configured, updated, and removed. These include system programs and application programs.

Without software, a computer does nothing. When you power on your computer, it first goes through an internally stored *booting process* (Section 4.11), which checks hardware and loads the operating system, among other things.

To perform any task, there must be a program specifying the step-by-step procedure for doing it. The power and usefulness of a computer mostly depend on the richness of software available for it.

The most important and crucial piece of software for a computer is its *operating system* (OS) (Section 4.1). Every general-purpose computer needs an OS to control and manage all hardware resources as well as all application programs. The *kernel* is the central part of an OS that stands between the hardware and all other programs. The OS interfaces to hardware on one side and to application software on the other (Figure 1.7). The *file system* is another important part of any OS and provides persistent storage of data in named files organized into a searchable hierarchy.

For most users, the same hardware under a different OS becomes a different computer. And, the same OS running on different hardware works essentially the same way. As depicted by Figure 1.7, users perform tasks on a computer by running specific applications that interface to the hardware through the OS. Each new application installed gives a computer new abilities. Well-known applications help you browse the Web, send/receive email, listen to music, watch movies and video, make voice/video calls, chat online, prepare tax returns, prepare documents and presentations, work with spreadsheets, and more.

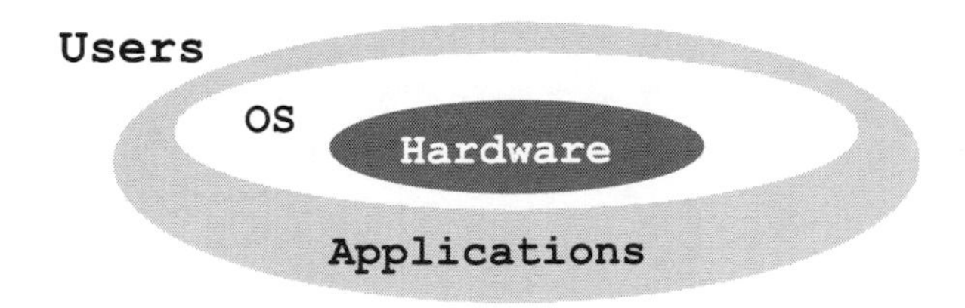

FIGURE 1.7 Operating System

1.5 Programming

Computers are all-purpose machines. A computer can be "told" to do one task or another, and it will perform the task according to step-by-step instructions. The activity of creating and refining the instructions for performing different tasks is computer programming.

A computer program is a complete set of procedures for performing particular computations or tasks. Each procedure is specified with a sequence of instructions. A program is written according to a precise format, called a *programming language*. The programming language defines the *syntax* (grammar rules) and *semantics* (meaning) of allowed constructs. When writing a program, any deviation from the language will usually result in a *syntax error*. Many kinds of errors, called *bugs*, are possible in programs. *Debugging* refers to the activity of testing and correcting program errors.

A computer understands a built-in *machine language* (Section 10.5), using only zeros and ones, and can execute instructions specified only in that machine language. As you can imagine, writing code in machine language is very difficult. An *assembly language* allows symbolic names for instructions and variables, making it somewhat easier. *High-level languages* (Section 10.7) make program writing much easier. A high-level language program is translated into machine language (Figure 1.8) before it can be executed by a particular computer. A *compiler* is a software tool that does the required translation, known as compiling, for a given high-level language. Instead of compiling, some high-level languages may use an *interpreter*, another application, to carry out the language instructions.

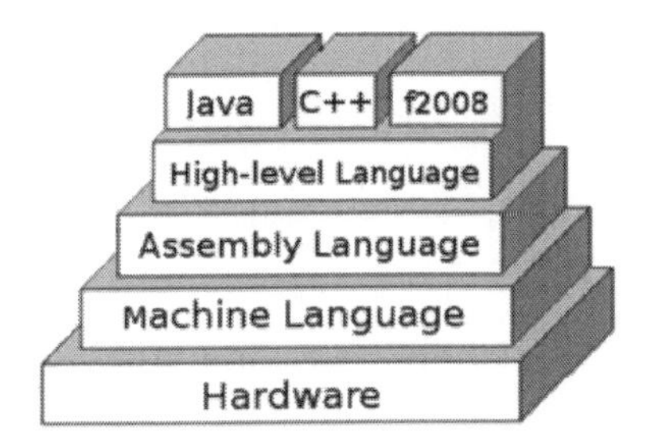

FIGURE 1.8 Programming Language Levels

In 1954, IBM (San Jose) developed FORTRAN, the first successful high-level language. It is especially suited for programming scientific and engineering applications. Evolved versions of it are still in use today (`f2008` is a new FORTRAN standard). Currently, widely used general-purpose languages include C/C++, Visual Basic, and Java. PHP and JavaScript, for Web programming, are examples of special-purpose languages.

1.6 Syntax and Semantics

Learning a programming language involves getting to know its syntax and semantics. Syntax deals with rules to correctly combine symbols in the language in a grammatically correct way. Semantics deals with the actual meaning of correct constructs. To illustrate their distinction, let's turn to a natural language, such as English. For example, the sentence, "Bees shed leaves in the autumn," has correct syntax but no valid semantics. In a programming language, the example expression

```
a = b + c
```

has correct syntax and has the semantics, "Add the values of variables `b` and `c` and assign the result as value of the variable `a` replacing any old value `a` may have." The same expression in mathematics becomes an equation and carries the very different meaning, "The values on both sides are equal." In mathematics, the same equality relation can be expressed as

```
(b + c) = a
```

But in a program, that expression would have incorrect syntax because the left-hand side of an assignment must be a variable.

Often, the semantics of the same expression can be different depending on the *context* of its usage. A well-known example is the English sentence, "Mary had a little lamb." It may refer to the food she had at lunch or the pet she had as a child. The idea that the same symbol or expression can mean different things in different contexts is important in real life as well as in computing. In a computer program, the symbol `0` may be a number or the logical constant **`false`**, and the symbol `1` may be a number or the logical constant **`true`**. And it is good practice to always check the context of something before jumping to any conclusions.

> **CT: Beware of Semantics**
>
> *The meaning of a word or symbol is defined by its language. The semantics of an expression often depends on its context.*

For example, in C/C++, the symbol `!` means "logical not," and the symbol `||` means "logical or."

1.7 Flowcharts

Programming involves creating step-by-step procedures (processes) suitable for execution on a computer. Such a procedure must leave nothing to chance. It specifies where to start, exactly what to do at each step, what is the next step, and when to stop. Programming requires watertight thinking, attention to details, and anticipating all possibilities at each step. Beginners find this difficult and tedious. But, thinking like a computer program can bring much advantage to us in almost all other activities.

A *flowchart* presents a procedure visually with words and diagrams. Creating a flowchart can help us design a procedure before writing it in a programming language. For any procedure, we can use a flowchart to plan the sequences of steps, to refine the solution logic, and to indicate how to handle different possibilities. Figure 1.9 shows a simple flowchart for the task of "getting up in the morning." We begin at the `Start` and follow the arrows to each

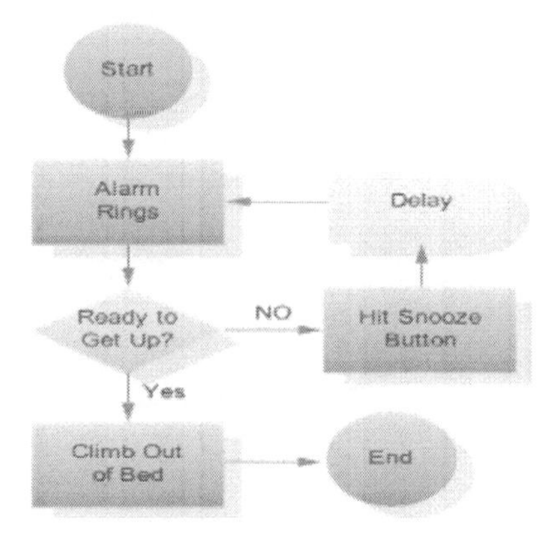

FIGURE 1.9 A Simple Flow Chart

next step. In programming, the flow from step to step is called the *control flow.* A diamond shape is used to indicate a fork in the path. Which way to turn depends on the conditions indicated. Obviously, we use diamond shapes to anticipate possibilities. The snooze option leads to a branch that repeats some steps. In programming, such a group of repeating steps is called a *loop.* The procedure ends when the person finally climbs out of bed.

As another example, let's look at a flowchart for troubleshooting a lamp (Figure 1.10). The very first step after `start` is significant. Although the

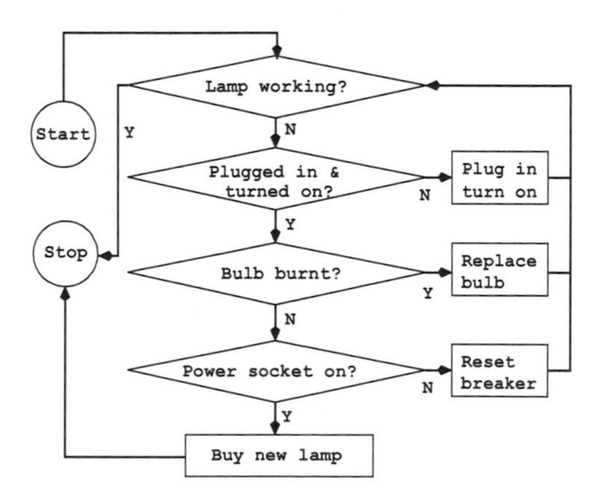

FIGURE 1.10 Lamp Fixing

purpose of the procedure is to troubleshoot a lamp, we, nonetheless, make no implicit assumption that the lamp is not working. Without this step at the beginning, the procedure would potentially troubleshoot a perfectly good lamp, and, worse yet, would decide to replace it with a new lamp!

> **CT: Ready for All Contingencies**
>
> *When executing a task, be prepared for all contingencies every step of the way.*

Each of the next three steps tests for a particular problem and makes a fix. Then the same procedure is reiterated by going back to step one to determine if the lamp is now working. This further demonstrates the importance of step one.

Steps 2, 3, and 4 are ordered according to their probabilities. That is, we check the most common problem first. It makes the procedure more efficient. For this procedure, it still works if we take these three steps in a different order. However, in general, step sequencing is important, and changing the order may break the procedure.

> **CT: First Things First**
>
> *Perform tasks in the correct order. Avoid putting the cart in front of the horse.*

Our third example flowchart (Figure 1.11) outlines a plan for a computer program that receives a series of words on a line and echos back the given words in reverse order. The word count `n` is a variable that is set to the number of words given. If no words are given, a contingency not to be ignored, or remain (i.e., `n` is zero), the program ends by displaying a NEWLINE character. Otherwise, it displays the `n`th word, sets the new word count to `n-1`, and loops back to test the value of `n`. Because `n` is reduced by 1 with each iteration, the procedure eventually will end.

> **CT: Check before Proceeding**
>
> *Always check before proceeding or repeating a task or a sequence of steps.*

Flowcharting is very useful in activity planing and working out systematic procedures for tasks. Try it yourself at the CT site (**Demo:** `DoFlowchart`).

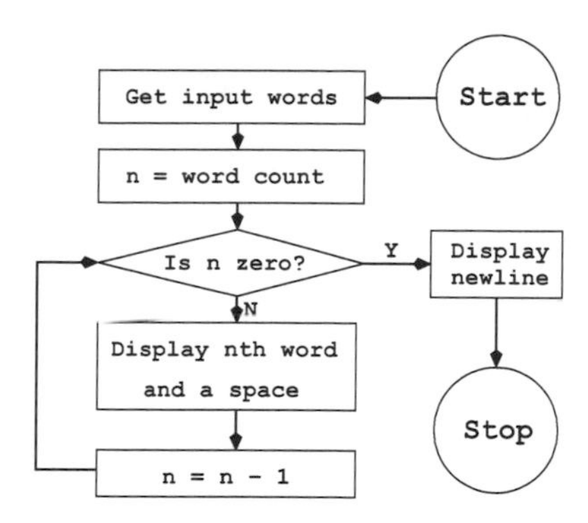

FIGURE 1.11 Reverse Echo

1.8 Algorithms

The origin of the word "algorithm" traces back to the surname Al-Khwārizmī of a Persian mathematician (780–850 CE), who was well-known for his work on algebra and arithmetic with Indian numbers (now known as Arabic numbers). The modern-day meaning of algorithm in mathematics and computer science relates to an effective step-by-step procedure to solve a given class of problems or to perform certain tasks or computations.

It is generally agreed that a procedure is an algorithm if it can be carried out successfully by a Turing machine (Section 1.2). Specifically, a procedure becomes an algorithm if it satisfies all of the following criteria[3]:

- Finiteness: The procedure consists of a finite number of steps and will always terminate in finite time.
- Definiteness: Each step is precisely, rigorously, and unambiguously specified.
- Input: The procedure receives a set of data, possibly empty, provided to the procedure before it starts. Possible values for the data may vary within limitations.
- Output: The procedure produces a nonempty set of results.
- Effectiveness: Each operation in the procedure is basic and clearly doable.

[3]Following criteria given by D. Knuth in his book, *The Art of Computer Programming, Vol. I.*

CT: Make It an Algorithm

Try to make a procedure or solution for any recurring task into an algorithm. You'll appreciate the resulting rigor, precision, robustness, and efficiency.

For a given problem, there usually are multiple algorithms for its solution. The design and analysis of algorithms are central to computer science and programming. The goal is to invent ever more efficient procedures to solve problems and perform computations. Let's look at a simple example. Say we have a long list of names and addresses, like those in a phone book, and we want to find someone on the list. What would be an algorithm to do that? This is the well-known and well-studied *searching* problem in computer science.

- Searching Algorithm A: This is the simple-minded, brute-force way. Start with the first entry on the list, then go through each successive entry until the desired entry is found or the list is exhausted.
- Searching Algorithm B: This is the *binary search* algorithm. It assumes that the list is in some kind of order (alphabetical for example). Start looking at the entry M at the midpoint of the list. We have three possibilities: (1) M is the entry we want, (2) M is after the entry we want, (3) M is before the entry we want. We stop if (1) is true. The list to search becomes the first half (case 2) or the second half (case 3) of the list. We repeat the same procedure (recursion) until the entry is found or the list is exhausted.

Figure 1.12 illustrates the binary search algorithm with a flowchart, where the list `L` is assumed to be already in increasing order, `L[0]` is the first entry on the list, and `L[n-1]` the last. Trace the algorithm on the flowchart, satisfy yourself that it works even if the list L is empty, or has just one entry, at the beginning. It is important for an algorithm to cover all input contingencies.

CT: Consider Extreme Cases

Always treat any extreme or trivial cases at the beginning of a procedure or loop.

Now, let's compare and analyze these two search algorithms. In each, the

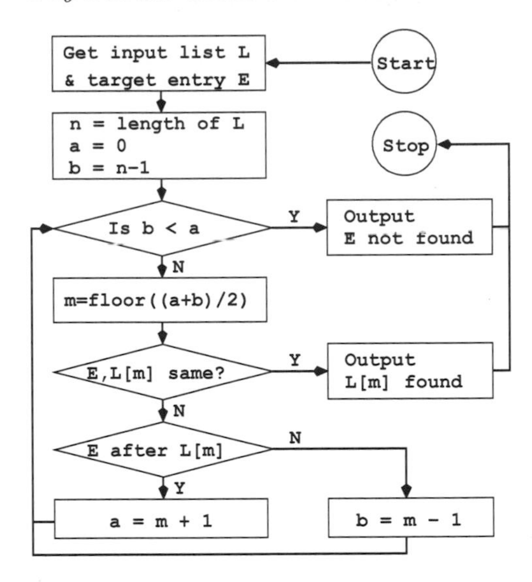

FIGURE 1.12 Binary Search Algorithm

basic operation is a comparison. Each comparison operation compares the target entry `E` with some entry on the list. For a list with n entries, the brute-force algorithm can take up to n comparisons. Binary search halves the list size with each comparison and therefore will take up to $\log_2(n)$ comparisons. If $n = 2^{32}$, we need at most 32 comparisons. What a big difference in efficiency! Granted that binary search requires first putting the list in order while the other algorithm does not. But ordering is a one-time cost; after that each search will be very fast.

Once an algorithm is chosen, programming can then produce an implementation of it written in a given programming language.

1.9 Pseudo Code

We see that a flowchart can help describe an algorithm in more precise ways. Ultimately, an algorithm can be implemented with computer programs that follow strict rules on acceptable constructs, punctuation, and so on.

But, one does not need a formal computer language to clearly describe an algorithm. Algorithms can be written in *pseudo code*—a clear set of step-by-step instructions written in a natural language, such as English.

For example, let's look at an algorithm for comparing two integers. The input consists of two arbitrary integers, and the output is `smaller`, `equal`, or `larger`, indicating the first number is less than, equal to, or bigger than the second number.

Here is our pseudo code:

Algorithm `compare`:
Input: Integer `a`, integer `b`
Output: Displays `smaller`, `equal`, or `greater`

1. If `a` is less than `b`, then output "`smaller`" and terminate

2. If `a` is equal to `b`, then output "`equal`" and terminate

3. Output "`larger`" and terminate

Note the instruction "terminate" ends the procedure and therefore stops the control flow from going to the next step.

Often, an algorithm will take the form of a *function* instead of a procedure. A function is a procedure that returns a value computed from its input. For example, the pseudo code

Algorithm `abs`:
Input: Any number `x`
Output: Returns absolute value of `x`

1. If `x` is greater than zero, then return `x`

2. Return `-x`

describes an algorithm to compute the absolute value function. Note `return x` means stop and produce the value `x`.

As another example, let's look at the factorial function. For non-negative n, the notation $n!$ (n factorial) is the product $n \times n-1 \times n-2 \times ... \times 1$. 0! is defined to be 1. Here is the factorial algorithm.

Algorithm `factorial`:
Input: Non-negative integer `n`
Output: Returns $n!$

1. Set `ans = 1`

2. If `n equals 0`, then return `ans`

3. If `n equals 1`, then return `ans`

4. Set `ans = ans` $\times$ `n`

5. Set `n = n-1`

6. Go to step 2

CT: Step by Step

Start with the input (known data). Plan a sequence of steps toward the goal (desired result). Take care of all contingencies at each step.

1.10 The Euclidean GCD Algorithm

Most of us first learned about the *greatest common divisor* (GCD) of two integers in middle school. For any two integers a and b, not both zero, $GCD(a, b)$ is the largest integer that divides both A and B evenly.

In middle school, the GCD is often referred to as the *greatest common factor* (GCF). And we were taught to find all factors of a and all factors of b and then figure out the greatest common factor from there. This method, while providing some insight into numbers and their factors, is terribly inefficient.

The most efficient method, attributed to the famous Greek mathematician Euclid (300 BCE), is the Euclidean algorithm, which is based on the fact

$GCD(a, b) = GCD(a, a - b).$

Here is a modern version of it.

Algorithm `EuclideanGCD`:
Input: Two integers `a` and `b`
Output: Returns gcd(`a`,`b`)

1. If both `a=0` and `b=0`, then refuse to continue and terminate
2. If `a < 0`, then set `a = (-a)`
3. If `b < 0`, then set `b = (-b)`
4. If `b = 0`, then return `a`
5. Set `r` = remainder of `a` divided by `b`
6. Set `a = b`
7. Set `b = r`
8. Go to step 4

$GCD(a, b)$ is undefined if both a and b are zero. If a or b is negative, we can use its absolute value instead. These situations are taken care of in the first three steps. Then, we execute steps 4 through 8 repeatedly until r becomes zero. To illustrate how well it works, let's trace the algorithm with input $a = 546$, $b = 1610$ (Table 1.1).

TABLE 1.1 Tracing the Euclidean GCD Algorithm

a	b	r
546	1610	546
1610	546	518
546	518	28
518	28	14
28	14	0

Thus, we get $GCD(546, 1610) = GCD(28, 14)$, which is 14. The result is obtained with four divide and take remainder operations.

Actually, you don't need a table to trace this algorithm. Just list the sequence of remainders produced, like this:

```
546  1610  546  518  28  14  0
```

Trace some examples yourself. You can use the **Demo:** `InteractiveGCD` tool at the CT site.

The Euclidean algorithm also demonstrates the seemingly obvious fact that knowledge in the problem domain, mathematics, in this case, can help create efficient ways to solve a problem, the GCD, in this case.

> **CT: Apply Domain Knowledge**
>
> *Expert knowledge in the problem domain can be very helpful in devising efficient and effective solutions.*

On the other hand, insufficient domain knowledge can cause major problems. The US `healthcare.gov` website debacle in late 2013 is a case in point. The simple website design principle of separating window shopping from the checkout process, among other obvious software engineering and web design features, was absent.

1.11 Goals and How to Get There

The preceding algorithms deal with simple tasks. But we can use the same step-by-step method to solve more complicated problems and achieve well-defined goals. Each step in a complicated algorithm can achieve a subtask needed in the overall solution. Each subtask is done by its own algorithm. Thus, the solution becomes a sequence of *calls* to procedures or functions for subtasks. If a subtask is still complicated, it can be broken down the same way again.

As an example, let's consider checking any given email address for correct form. An algorithm for achieving this goal can be as follows.

Algorithm `emailCheck`:
Input: An email address `em`
Output: Returns `true` or `false`

1. Does `em` have the form *user*`@`*host* where neither *user* nor *host* may contain the character `@`? If not, then return `false`.

2. Is *user* either one or more words separated by dots or several words separated by spaces and enclosed in quotes (`" ... "`)? If not, then return `false`.

3. Is *host* a correct domain name or IP address? If not, then return `false`.

4. Return `true`.

If the given email is not in correct form, the algorithm returns `false`. Otherwise, it returns `true`. With this top-level breakdown, we can then create algorithms for each of the subtasks in steps 1–3.

> **CT: BREAK IT DOWN**
>
> *Solve a complicated problem by breaking it down into a sequence of smaller subproblems. Each subproblem can be broken up the same way. Eventually, the the subproblems will become easy to solve.*

1.12 Road Crossing

Why did the chicken cross the road?

To get to the other side!

FIGURE 1.13 Why?

For us, we also have reasons to cross the road. On one side, we have fussy, vague, confused, wishful, emotional, impulsive, optimistic, or pessimistic thinking. We want to get to the other side and learn computational thinking.

In this chapter, we see the algorithm side of CT, where we have well-defined starting and end points, carefully arranged sequencing, provisions for all implications and contingencies at every step, and a way to break down complicated tasks into simple and doable ones.

Take driving a car as an example. What is the goal? It is to get to a destination safely. It is not enjoying the sound system, watching the scenery, or engaging in conversation, although we have nothing against any of that as long as it does not get in the way of safe driving.

Stopped at a traffic light, we wait for the light to turn green. But, we may need to run the red light if an 18-wheeler is about to crash into us from behind. That means we need to be checking our rear-view mirror while waiting for the green light. When the light turns green, do we blindly rush into the intersection? What if a car is running the tail end of the yellow light or the red light? Thus, the goal is not to obey traffic signals, but to make sure it is safe. In the United States, a car crash kills a person every 12 minutes on average. If you are thinking straight, is a car a fun machine or a dangerous one? CT can keep you focused on the goals, make you pay attention to details, plan for contingencies, and shield you from distractions. CT can save the day, and perhaps even your life!

Now let's apply CT to the task of "getting ready to drive a car" and write down an algorithm-inspired predrive checklist.

Algorithm `predriveCheck`:

1. Am I ready to leave? Forgot to bring anything?

2. Walk around the car, check windows, tires, lights, back seat, and any objects near the car.

3. Get in the car, foot on brake, close and lock all doors.

4. Adjust seat and steering column positions as needed. Check positions of all rear-view mirrors, buckle up.

5. Check the instrumentation panel, pay attention to the fuel level.

6. If necessary, familiarize yourself with the controls for lights, turn signals, wipers, heat/AC, and emergency signal. Make sure they are working properly.

7. Release the hand break, start the engine, shift gear.

8. Make sure the gear is in D or R as intended, then start driving.

Airlines have developed rigorous preflight checklists for safety. Incidents, sometimes fatal, happen when pilots and crew, failing computational thinking, do not follow the exact procedure. In the United States, parents neglecting to engage safety locks on firearms cause too many unfortunate injuries and deaths for children.

Exercises

1.1. What is a CPU? Name a few (three) modern CPUs and their speeds.

1.2. What is a Turing machine? What does computable mean?

1.3. Explain in your own words the concept of "internal state" of a machine.

1.4. What is syntax? Semantics? What's the difference. Also give examples.

1.5. What is computer memory? RAM? Hard disk? What is the difference?

1.6. What is the one most important feature that distinguishes a computer from other machines?

1.7. What is hardware? Software? What's the difference?

1.8. Create a flowchart for "Going to bed."

1.9. Set a favorite recipe of yours in pseudo code form.

1.10. What is an algorithm? What conditions must it satisfy to qualify?

1.11. Follow the Euclidean algorithm to compute gcd(136500, 1227655).

1.12. **Computize**: What are "the five Ws and one H"? How do they relate to CT?

1.13. **Computize**: Create a flowchart for "boiling water"? Pay attention to safety and consider all kinds of possibilities.

1.14. **Computize**: Whenever you are ready to leave from one place to another, what CT should apply? Please explain and give examples.

1.15. **Computize**: When reading or listening to political or commercial messages, what CT should apply? Please explain and give examples.

1.16. **Group discussion topic**: *According to the US Department of Transportation, in 2013, 32,719 people died in traffic crashes. That is almost 90 deaths per day.*

1.17. **Group discussion topic**: *When something goes wrong, how best to troubleshoot.*

1.18. **Group discussion topic**: *Smartphone camera and video settings and options. Importance of domain knowledge.*

Chapter 2

Bits, Bytes, and Words

Computers store and process digital data. Numbers, text, images, sound, and video must be stored in memory before computers can process them.

We shall explain the organization of computer memory and how numbers, characters, and other data are represented with bit patterns. Binary numbers, modular arithmetic, US-ASCII, and Unicode character encoding will be covered.

2.1 Digital Computers

Modern computers are *digital* because they store and process *discrete* rather than *continuous* information.

- Discrete data—Data are *discrete* when only certain distinct separate values are allowed. The number of chickens, letter grades, basketball scores, age, income, integers, fractions, and so on are examples. Discrete values have gaps in them.
- Continuous data—Data are *continuous* when all values in a continuous range, finite or infinite, are allowed. Length, weight, volume, temperature, pressure, speed, brightness, and so on are examples. Continuous values have no gaps separating them. Thus, even in a small range, there are an infinite number of continuous values.

In the past, analog computers processed continuous electronic waves. Such analog signals are hard to store, transmit, or reproduce precisely. In contrast, digital computers use integers to represent information and therefore avoid these critical problems. A continuous value, that of a sound wave, for example, can be *digitized* by sampling values at a number of discrete points (Figure 2.1). With enough sampling points, the continuous sound wave can be recreated. For example, a continuous sine wave $f(x) = sin(x)$ has an infinite number of continuous values between $x = 0$ and $x = \pi$. To digitize the sine function, we can pick points $\delta = \pi/1000$ apart and use values at 1001 discrete points: $sin(0)$, $sin(\delta)$, $sin(2 * \delta)$, $sin(3 * \delta)$, and so on all the way to $sin(\pi)$.

Information, represented in digital form, must be stored in computer memory to be processed. The most basic memory unit is a bit, which can represent

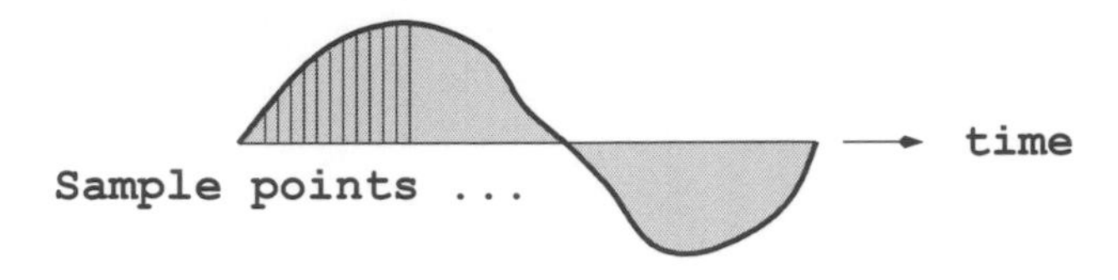

FIGURE 2.1 Sampling a Continuous Wave

either a `1` or a `0`. A *byte* is a group of 8 bits. A *word* consists of several (usually 4 or 8) bytes (Figure 2.2). Note in computing, as shown in Figure 2.2, we count starting from zero.

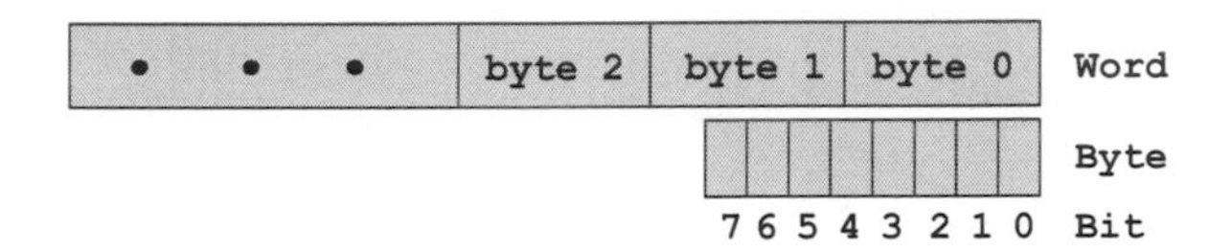

FIGURE 2.2 Bit, Byte, and Word

In a computer, the *Central Processing Unit* (CPU) is where logic computation on data takes place. The CPU loads data from memory in order to process them, then stores the results back to memory. A *word* is normally the smallest integral data entity a CPU is designed to manipulate. The size of the word, measured in bits, is determined by the size of CPU registers. A CPU register is able to load from and store to cache/RAM (*Random Access Memory*) all its bits as a unit at once. The word size is one of the most important hardware architecture features of a computer. For most modern computers, the word size is either 32 (4 bytes) or 64 (8 bytes).

For modern general purpose computers, the entire memory is an array of bytes (Figure 2.3), each of which can be *addressed* directly (called byte addressing), hence the term random access memory. A word of length n can hold memory addresses ranging from 0 to at most $2^n - 1$. Assuming each address has a byte, then the memory can have at most 2^n bytes. The actual allowable size of your RAM is often well below this theoretic upper bound due to other hardware limitations.

Digital information must be encoded by zeros and ones and stored in memory in order to be processed. Information stored in RAM is *volatile* and will disappear when the system is turned off. Nonvolatile or *persistent* data storage is provided by one or more storage drives, including hard-disk drives (HHD) or solid-state drives (SSD).

In computer product specifications, available memory in cache, RAM, or data storage is given in byte units:

- kilobyte (KB) = 1024 bytes
- megabyte (MB) = 1024 KB

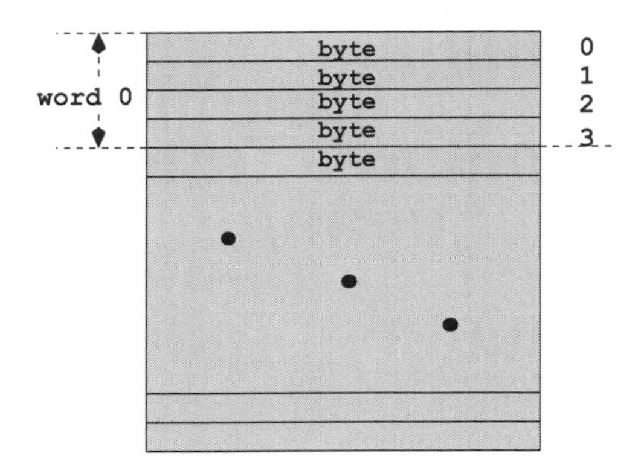

FIGURE 2.3 Memory Array of Bytes

- gigabyte (GB) = 1024 MB
- terabyte (TB) = 1024 GB
- petabyte (PB) = 1024 TB

Typical computer RAM sizes range from 1GB to 16GB.

Ordinarily, K (kilo) is a prefix for 1000 (in the metric system), but digitally (in the binary system), K is 1024. Similarly, as far as memory or data sizes go, M (Mega) is KK, Giga is KM, Tera is KG, and Peta is KT, keeping in mind always that K is 1024[1]

Bit Patterns

The basic way to represent data in computer memory is to use *bit patterns*. A bit pattern is a particular sequence of zeros and ones presented in a fixed number of bits. Table 2.1 shows all possible patterns made up of three bits. Each bit has two variations, 0 and 1. For each value of bit 1, there are two values for bit 0, resulting in $2 \times 2 = 2^2$ two-bit patterns. Similarly, for each value of bit 2, there are 4 patterns for bits 0 and 1, giving $2 \times 2^2 = 2^3$ three-bit patterns.

In general, the total number of different patterns with n bits is 2^n. Therefore, a byte can give you $2^8 = 256$ bit patterns, a 32-bit word $2^{32} = 4294967296$ bit patterns, and a 64-bit word $2^{64} = 18446744073709551616$ bit patterns.

In a digital computer, bit patterns are the only way to represent data. And, the same bit patterns can be used to represent different types of data, such as a number, a character, or a network address. We will see how numbers are represented next.

[1]In actual usage, there is often confusion between the metric and binary interpretations.

TABLE 2.1 The Eight 3-Bit Patterns

Bit 2	Bit 1	Bit 0
0	0	0
0	0	1
0	1	0
0	1	1
1	0	0
1	0	1
1	1	0
1	1	1

2.2 Binary Numbers

To compute with numbers, we need to represent them in memory. For integers, the *binary number* system uses only zeros and ones and is therefore particularly suited to be stored as bit patterns.

Numerals

Numerals are symbols we use to write down numbers. The numerals we use today (0, 1, 2, 3, 4, 5, 6, 7, 8, 9) are *Arabic numerals* derived from the Hindu-Arabic numeral system. Different civilizations have invented and used their own symbols for numbers. Figure 2.4 shows the Roman and Egyptian numerals.

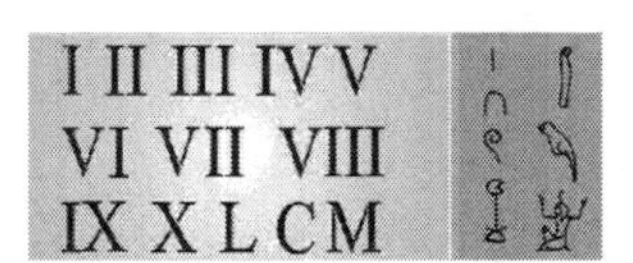

FIGURE 2.4 Roman and Egyptian Numerals

And Figure 2.5 shows ancient Chinese numerals. Simplified modern versions of these are still in use today.

With the digits, the Arabic numerals, we can write down numbers from zero to nine. We still need a way to represent larger numbers. We can invent new symbols, of course. Figure 2.6 shows some modern Chinese symbols for larger numbers.

How will we create new symbols to keep up with ever larger numbers? A systematic way must be found. An ingenuous solution is the *place value system*. For example, with three digits side by side, we can write down numbers up to, but not including, a thousand. Thus, the notation 379 means three hundreds, seven tens, and nine ones, or

$379 = 3 \times 10^2 + 7 \times 10 + 9 \times 1$

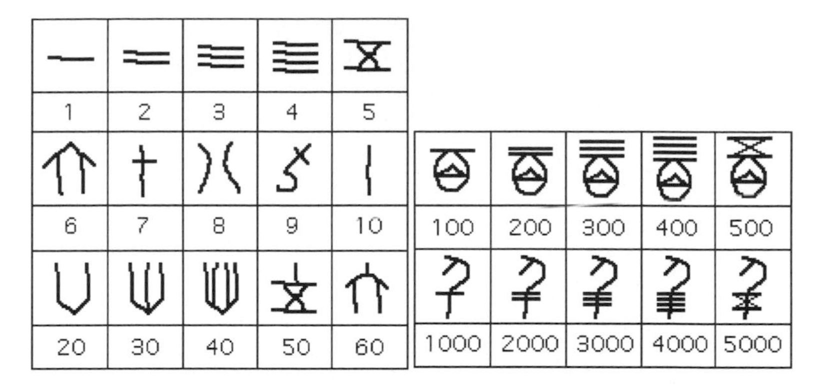

FIGURE 2.5 Ancient Chinese Numerals

十	百	千	万
ten	hundred	thousand	ten thousand

FIGURE 2.6 Larger Chinese Numerals

With the place value system, larger numbers simply require more places. Thus, we have a system for creating new symbols for numbers by combining the numerals. Such numbers, where each place represents a power of 10, are known as *base-10 numbers* or *decimal numbers.*

Base-2 Numbers

Binary numbers also use a place value system, just like decimal numbers, except the base is 2, not 10. We use only two numerals, 0 and 1, for binary numbers. Using one place, only two numbers can be represented, namely 0 and 1. To represent two in binary, we need to go to the next place. Thus, the binary number 10 means two

$1 \times 2 + 0 \times 1$

and the binary number 11 means three

$1 \times 2 + 1 \times 1$

Similarly, 101 means five

$1 \times 2^2 + 0 \times 2 + 1 \times 1$

With four bits, we can represent numbers from zero to fifteen (Table 2.2).

Keep in mind that, just as in decimal numbers, the least significant digit (bit) has the rightmost position and the most significant bit has the leftmost position. And the bit and byte positions are always counted from the right.

TABLE 2.2 Four-Bit Binary Numbers

Binary	Decimal	Binary	Decimal	Binary	Decimal	Binary	Decimal
0000	0	0001	1	0010	2	0011	3
0100	4	0101	5	0110	6	0111	7
1000	8	1001	9	1010	10	1011	11
1100	12	1101	13	1110	14	1111	15

To get familiar with binary notation, take any binary number in Table 2.2 and add 1 to it, carrying to the next higher position as you would in doing addition, you will get the bit pattern for the next binary number. Also visit these interactive demos on the CT website: **Demo:** `UpCounter` and **Demo:** `DownCounter`.

It is not surprising if you find binary numbers confusing at first. After all, as far as numbers, the decimal system is our mother tongue and is firmly ingrained in our thinking. Look at Figure 2.7 and see how you feel yourself. Binary numbers would become easier if we memorize the powers of 2 as well

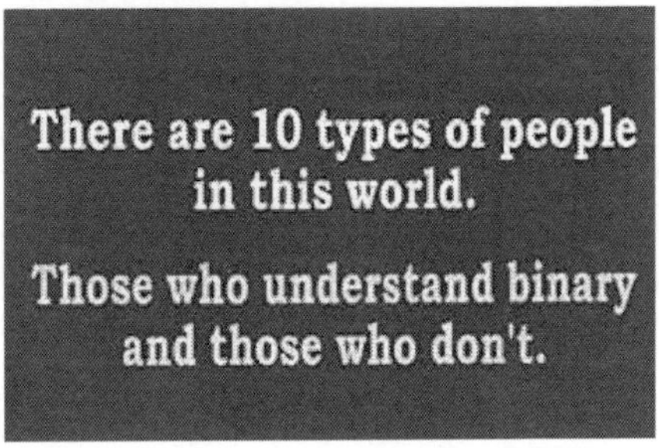

FIGURE 2.7 Popular T-shirt Design

as we do powers of 10. Here goes: 2 (bit 2), 4 (bit 3), 8 (bit 4), 16 (bit 5), 32 (bit 6), 64 (bit 7), 128 (bit 8), 256 (bit 9), 512 (bit 10), 1024 (bit 11), 2048 (bit 12), and so on.

Inside a digital computer, numbers are naturally in binary, and hardware support is provided for their operations, as long as they fit in a single word. A 32- or 64-bit word can represent numbers from 0 to $2^{32} - 1$ or $2^{64} - 1$. That's a lot of numbers. Still, it is far from covering all numbers. To handle larger numbers, we can use multiple words and write software for manipulating them.

> **CT: Meaning of Symbols**
>
> *Anyone can invent a symbol and assign it a meaning. The same symbol may have another meaning in a different context. Do not guess a symbol's meaning by its appearance. Always refer to its definition within the intended context.*

Numerals and numbers from different cultures, as well as the place value system, are examples.

Ancient Chinese Binary Symbols

I-Ching (易經), a Chinese text that traces back to the 3rd to the 2nd millennium BCE, introduced two symbols, yin (- -) and yang (—). Combinations of three or six yin-yang symbols form the 8 trigrams (八卦 Figure 2.8) or the 64 hexagrams (六十四卦). Thus, the concept of repeating two symbols to form

FIGURE 2.8 I-Ching Trigrams (八卦)

increasingly more symbols is an ancient one.

2.2.1 Numbers in Other Bases

While we are at it, why don't we look at numbers in other bases? Octal numbers are base 8 and use digits 0–7. Hexadecimal (hex) numbers are base 16 and use digits 0–9 followed by A (ten), B (eleven), C (twelve), D (thirteen), E (fourteen), and F (fifteen). The symbol 10 stands for eight in octal and sixteen in hexadecimal. The symbol 25 means

$2 \times 8 + 5$ (21 in octal)
$2 \times 16 + 5$ (37 in hex)

For interactive demos, visit the CT website (**Demo:** `OctalCounter` and **Demo:** `HexCounter`).

TABLE 2.3 Numbers in Different Bases

Decimal	11	17	23	29
Octal	13	21	27	35
Hex	B	11	17	1D
Binary	01011	10001	10111	11101

Three bits are needed for each octal digit and four bits for each hexadecimal digit. A byte of all 1s can represent the hex number FF (two hundred fifty-five). In general, we can use numbers in any base we desire, and we are not, limited by having two hands with ten digits.

> **CT: Evaluate Different Options**
>
> *No feature in a system should be treated as rigid or sacred. Ask "what if it is changed?" and you will begin to think flexibly, acquire a deeper understanding, and, perhaps, even make a breakthrough.*

Go to the CT website to experiment with counters where you can set the base (**Demo:** `ArbCounter`).

Octal and hex numbers are often used as a shorthand for the bit patterns representing them. Each octal digit specifies a 3-bit pattern, and each hex digit specifies a 4-bit pattern. Also, distinct prefixes, `0x` for hex and `0` for octal, are used to denote such numbers. This is especially true in programming languages. Table 2.4 shows the bits for each digit of `01357` or `0x2EF`.

TABLE 2.4 Octal and Hex Bit Patterns

Octal	1			3			5			7		
Binary	0	0	1	0	1	1	1	0	1	1	1	1
Hex	2				E				F			

Mixed-Base Numbers

Decimal numbers use powers of ten as place values; binary numbers powers of two; octal numbers powers of eight; and hex numbers powers of sixteen. Talking about flexible thinking, what about asking the question why the place values must always be a power of some fixed number?

Well, there is no reason for that at all. In fact, we can use any desired value for each place. Such numbers are called *mixed-radix* or *mixed-base* numbers. For instance, how about numbers with place values 1, 60, 60, 24, 7? Would that be crazy?

Actually, we use such numbers every day, literally. We have 60 seconds in a minute, 60 minutes in an hour, 24 hours in a day, and 7 days in a week, don't we? And we have 12 inches in a foot, 3 feet in a yard, and 1760 yards in a mile. Granted, such length measurements should have long ago been replaced by the metric system.

Applying abstraction (CT: ABSTRACT AWAY, Section 1.2), we realize that the place-value numbering system's essence, its intrinsic nature, is "a preassigned value for each place," and nothing more.

2.3 Positive and Negative Integers

Using binary notation, a word with n bits can represent numbers 0 to $2^n - 1$. These are known as *unsigned numbers*. But, we most definitely also need to work with signed numbers. If an integer is positive, we don't have a problem representing it. But what about negative integers?

An obvious idea is to use the highest bit in a word as a *sign bit* and use the rest of the word for the magnitude of the integer. If the highest bit is `1`, then it is a negative number. If it is `0`, a positive number. To make it easier to visualize, let's pretend that the word size is 3:

```
000 (is 0)  001 (is  1)  010 (is  2)  011 (is  3)
100 (is 0)  101 (is -1)  110 (is -2)  111 (is -3)
```

With sign and magnitude, using n bits, we can represent $2^{n-1} - 1$ positive numbers, $2^{n-1} - 1$ negative numbers, and zero twice.

The sign and magnitude method does not work as well as the *two's complement* method, which is widely used by computers. From the binary representation of an integer m, we use the following algorithm to find the binary representation of $-m$.

Algorithm `2sComplement`:
Input: An integer `m`
Output: Returns `-m`

1. Obtain the bit pattern for `m`

2. Obtain a new binary number by flipping each bit in `m` (switch `0` to `1`, and `1` to `0`)

3. Add `1` to the resulting number and return it as value

Again, for $n = 3$, we have the following 8 numbers

```
000 (is 0)  001 (is  1)  010 (is  2)  011 (is  3)  100 (is 4 or -4)
000 (is 0)  111 (is -1)  110 (is -2)  101 (is -3)  100 (is 4 or -4)
```

Figure 2.9 shows an application of the two's complement algorithm with an 8-bit word.

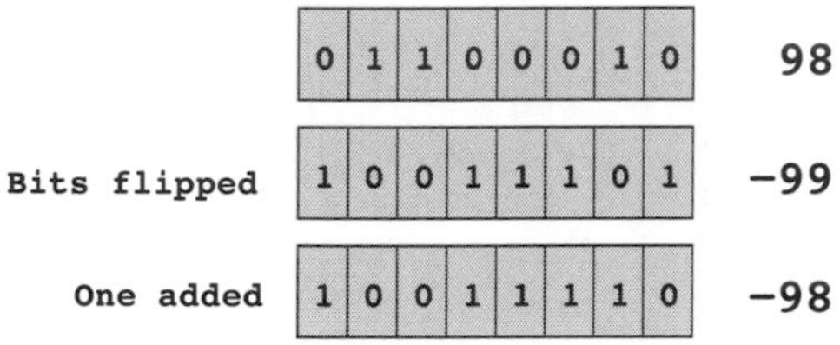

	0	1	1	0	0	0	1	0	98
Bits flipped	1	0	0	1	1	1	0	1	−99
One added	1	0	0	1	1	1	1	0	−98

FIGURE 2.9 Obtaining Negative 98

Note the following.

- The same algorithm turns a positive number into a negative number and vice versa.

- There is only one representation for zero.

- Negative numbers have a leading (left-most) one bit, and positive numbers have a leading zero bit, except 2^{n-1}, whose positive and negative representations turn out to be the same. Theoretically, this number is the most negative and most positive number possible for an n-bit word. But, most programming languages treat this number as negative. And define the largest positive number to be $2^{n-1} - 1$.

- Adding the representation of m and $-m$ results in zero, discarding any carry to the next, nonexistent, higher bit position (Section 2.4).

Try adding 98 and -98 in Figure 2.9 yourself.

Using the two's complement system for negative numbers has a very important advantage. Namely, the fundamental arithmetic operations of addition, subtraction, and multiplication (Section 2.4) do not need to check the sign of the number and can proceed as if all numbers are non-negative (**Demo: `SignedCounter`**). For division, the simplest way is to convert to positive, perform the division, and adjust the signs of the quotient and remainder later. Further, these procedures don't need changing when working with an increased word length.

2.4 Modular Arithmetic

As mentioned before, a word of size n can represent 2^n numbers, zero to $2^n - 1$ for unsigned integers and -2^{n-1} to 2^{n-1} for signed integers. CPU-supported arithmetic on such word-size numbers (addition, subtraction, multiplication, and integer division[2]) must produce results stored within a single word. To work within the limit imposed by the word length, *modular arithmetic* is used. We will describe modular arithmetic, a mathematical concept, presently.

In modular arithmetic, we limit the size of possible integers by a positive integer called the *modulus*. If m is the modulus, then the possible integers range from 0 to $m - 1$. With modular arithmetic, when a result is out of range, we add or subtract a multiple of m to bring the result in range. Let's see how this is done.

Definition: *For modulus m and two arbitrary integers a and b, a is congruent to b modulo m if m divides $(a - b)$ evenly.*

In other words, a is congruent to b modulo m if the remainder of $a - b$ divided by the modulus is zero. We use

$$a \equiv b \mod m$$

to denote that a is congruent to b modulo m. Let's look at an example where $m = 2$.

$$\ldots \equiv 8 \equiv 6 \equiv 4 \equiv 2 \equiv 0 \equiv -2 \equiv \ldots \mod 2$$
$$\ldots \equiv 7 \equiv 5 \equiv 3 \equiv 1 \equiv -1 \equiv -3 \equiv \ldots \mod 2$$

Thus, the set of all integers is divided into two disjoint subsets: the even and the odd numbers. All members within each subset are congruent mod 2. Each subset is called a *congruence class mod* 2. Under the modulus 2, we can use 0 to represent any even number and 1 to represent any odd number.

Now for modulus 16, we have

$$\ldots \equiv 48 \equiv 32 \equiv 16 \equiv 0 \equiv -16 \equiv -32 \equiv \ldots \mod 16$$
$$\ldots \equiv 49 \equiv 33 \equiv 17 \equiv 1 \equiv -15 \equiv -31 \equiv \ldots \mod 16$$
$$\ldots \equiv 50 \equiv 34 \equiv 18 \equiv 2 \equiv -14 \equiv -30 \equiv \ldots \mod 16$$

and so on for a total of 16 congruent classes.

Use of modular arithmetic can be found on the familiar watch face (Figure 2.10), where we have a limit of twelve hour marks. We count from 0 to the 11th hour, then start to reuse the marks again. This is why we need to use AM and PM to distinguish which hour it is out of the 24 hours in a day.

Modular congruence is an *equivalence relation*, because it satisfies all of the following three conditions.

For any integers a, b, c and modulus $m > 0$

- Reflexivity: $a \equiv a \mod m$ (Proof: $a - a = 0$, which is divisible by m.)

[2]Integer division can produce an integer quotient or an integer remainder.

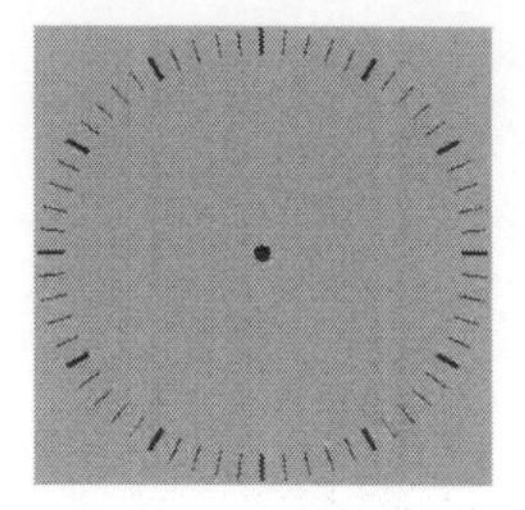

FIGURE 2.10 Watch Face

- Symmetry: $a \equiv b \mod m$ implies $b \equiv a \mod m$ (Proof: If m divides $(a - b)$, then it divides $(b - a)$.)

- Transitivity: $a \equiv b \mod m$ and $b \equiv c \mod m$ implies $a \equiv c \mod m$ (Proof: m divides $(a - b) - (b - c)$ because it divides both terms.)

Computationally, "mod m" simply means "divide by m and take remainder." For an expression *exp* involving addition, subtraction, and multiplication, *exp* mod m can be computed by performing mod m whenever necessary to reduce the size of the operands and intermediate results. The final result will always be the same. This way, we keep the numbers under computation always within 0 through $m - 1$. There is no overflow problem for division. To compute a/b, the fact that both a and b are already correctly represented means their quotient and remainder must also be in the correct range.

Computer hardware supports modular arithmetic with a modulus $m = 2^n$, where n is the word size (32 or 64, for example). This is done simply by ignoring any overflow to the left of the nth bit. Using modular arithmetic meets the need to work within the limit imposed by the word length.

Note that if a sequence of arithmetic operations on integers results in a final answer within the modular range, then using modular arithmetic for all those operations will result in the same answer. This happens even if some intermediate result may cause overflow.

For simplicity, consider numbers in 4-bit words. The allowable bit patterns range from 0 to 15. The arithmetic computation $3 \times 7 - 4 \times 5$, when carried out this way,

$3 \times 7 \mod 16$ producing 5
$4 \times 5 \mod 16$ producing 4
$5 - 4 \mod 16$ producing 1

gives you the correct final answer, namely 1.

CT: MIND RESOURCE LIMITATIONS

Be mindful of resource limitations. Find ways to operate within such limitations. Use resources wisely. Guard against running out of resources.

Keeping this in mind may help you avoid running out of gas on a trip or motivate you to conserve, reuse, and recycle materials.

In many programming languages, C/C++ and Java included, arithmetic operations (`+`, `-`, `*`, and `/`) are implicitly modular. And the operator `%` is used for an explicit mod operation:

`x % m` (divide `x` by `m` and take remainder).

Modular arithmetic is not only important for understanding arithmetic on binary numbers in a computer. It is also used widely in mathematics and extensively in modern cryptography.

CT: SYMBOLS CAN BE DECEIVING

Pay special attention to the same symbol when used in a different context. It may mean similar but different things. Such confusion may not cause problems immediately. That is dangerous.

Beware that the familiar mathematical symbols `+`, `-`, `*`, and `/` mean modular arithmetic in programming languages. Failure to guard against overflow in programs is a source of bugs hard to find. Such a program may work most of the time until some input situation causes an overflow.

When working with signed numbers, overflow occurs:

- When adding two positive numbers and getting a negative result.
- When adding two negative numbers and getting a positive result.
- When multiplying two numbers and getting the wrong sign in the result.

Such conditions are commonly disregarded in hardware, and programmers must detect these overflow conditions in their programs.

An actual example in advertising may help illustrate this CT concept.

An ad for drop down cloth hanger claims to save space in your closet by showing a picture of a closet full of clothing, side by side with a picture of the closet half empty using the product. The caption read, "The same closet using wonder hangers." When *Consumer Reports* pointed out that the pictures actually showed different numbers of clothing pieces, the company responded: "We only said the closet was the same." Ha ha ha, buyers beware.

2.5 Base Conversion

Now let's challenge ourselves to the task of finding the binary representation for any particular integer $a \geq 0$. Given any non-negative integer a and base b, our goal is to find the base b digits for a.

To illustrate our method, let's look at an example and think about it backwards.

$$a = 13 = 1 \times 2^3 + 1 \times 2^2 + 0 \times 2 + 1$$

The binary notation for $a = 13$ is `1101`, where $d_0 = 1$ (digit 0), $d_1 = 0$ (digit 1), $d_2 = 1$, and $d_3 = 1$.

$$a = 13 = d_3 \times 2^3 + d_2 \times 2^2 + d_1 \times 2 + d_0$$

If we divide both sides of the preceding equation by 2 and take remainder, we have

$$d_0 = remainder(a, 2)\,.$$

Now we set a to $(a - d_0)/2$, and get

$$a = d_3 \times 2^2 + d_2 \times 2 + d_1\,.$$

This leads to the value of d_1

$$d_1 = remainder(a, 2)\,.$$

Then we set a to $(a - d_1)/2$ and repeat the same steps until a becomes zero.

The general algorithm for base conversion can now be specified.

Algorithm `baseConversion`:
Input: Non-negative number to be converted a, the desired base b
Output: Displays sequence of digits for a in base b

1. Set $i = 0$
2. Set $d_i = remainder(a, b)$, set $a = quotient(a, b)$
3. If $a = 0$, then display base b digits d_i through d_0 and terminate
4. Set $i = i + 1$

5. Go to step 2

Call the algorithm with $b = 2$ for converting to binary, $b = 8$ to octal, and so on. Go to the CT website for an interactive version (**Demo:** `BaseConversion`).

> **CT: START FROM THE END**
>
> *It is natural to start from the beginning. But starting from the end can often be an effective problem-solving strategy.*

When deriving algorithm `baseConversion`, we assumed we had the desired digits, the end result (namely the d_is), already. Then, we see how each digit can be easily obtained by integer quotient and remainder operations.

In planning an event, such as a birthday party, a wedding, or an academic conference, it is best to plan for the "EVENT" day first, then what must be complete on EVENT −1 day or EVENT −1 week, and so on. Working backwards like this, we can sequence the order of tasks and allocate enough time for each task—and finally, when to start and what to do first!

In a sense, the starting and ending points are conceptually the same. For example, when taking a trip from home to destination A, we would produce a copy of the Google Maps directions. But, we may forget that we also need directions to come back home. If we computize, we would have directions from destination A back home printed at the same time. Actually, it would be more convenient for users if Google Maps automatically produces round-trip directions by default, or at least as an option[3].

Now, let's turn our attention to data representation for characters.

2.6 Characters

Number processing is fundamental, but computers need to handle other types of data, among which perhaps the most important is text or character data. Again, bit patterns are used to represent individual characters. And there are a few widely used standards for character code including US-ASCII and UNICODE.

2.6.1 US-ASCII

Basically, each character can be assigned a different binary number whose bit pattern represents that character. For example, the American Standard

[3]No such Google Maps provision as of Fall 2015.

Code for Information Interchange (US-ASCII) uses 7 bits (binary 0 to 127) to represent 128 characters: 0-9, A-Z, a-z, SPACE, CR, NEWLINE, punctuation marks, symbols, and control characters.

Table 2.5 shows the US-ASCII code (in decimal) for some common characters. Such tables are also widely available in octal and hex.

TABLE 2.5 US-ASCII for Common Characters

	30	40	50	60	70	80	90	100	110	120
0		(	2	<	F	P	Z	d	n	x
1		)	3	=	G	Q	[	e	o	y
2		*	4	>	H	R	\	f	p	z
3	!	+	5	?	I	S	]	g	q	{
4	"	,	6	@	J	T	^	h	r	|
5	#	-	7	A	K	U	_	i	s	}
6	$	.	8	B	L	V	`	j	t	~
7	%	/	9	C	M	W	a	k	u	DEL
8	&	0	:	D	N	X	b	l	v	
9	'	1	;	E	O	Y	c	m	w	

Thus, we have, for example, these character representations

```
'0'  00110000 (48)      '9'  00111001 (57)
'A'  01000001 (65)      'Z'  01011010 (90)
'a'  01100001 (97)      'z'  01111010 (122).
```

Note that the code for characters `'0'`–`'9'` are not the same as the binary number representations for 0–9 as numbers. The ASCII code for characters also results in an induced ordering of characters, known as a *collating sequence.* The ordering is based on the numerical value of each character's ASCII code. For example

```
'0'<'1'< ... <'9'<'A'<'B'< ... <'Z'<'a'<'b'< ... <'z'.
```

The collating sequence makes it easy to arrange texts, such as a list of names, into a standard order. A collating sequences may facilitate arranging text into an alphabetical order, but that may not be the same as the traditional ordering used in dictionaries.

US-ASCII is a subset of ISO-8859-1, a standard consisting of 191 eight-bit encoded characters commonly used in Western languages.

2.6.2 Unicode

Unicode is an international standard for encoding, representation, and handling of text data from most of the world's writing systems. It now contains more than 110,000 characters from 100 languages/scripts. The Unicode Consortium, an international collaboration, publishes and updates the Unicode standard.

Unicode allows the use of characters from all of the covered languages, which is practically all languages known to men, in a single written document. This is very advantageous, especially in a world increasingly interconnected by the Internet and the Web. The success of Unicode and its widespread use brought about standards on software *internationalization and localization.*

Unicode uses hexadecimal (hex) `0` through `10FFFF`, a total of 1,114,112 *code points*, to represent character symbols. Each code point is denoted `U+`, followed by 4 to 6 hex digits (Table 2.4, Section 2.2.1). As an example, here is the author's Chinese name with Unicode in hex.

`U+738B` 王 `U+58eb` 士 `U+5f18` 弘

Among different Unicode encoding formats, UTF-8 is the most widely used. The **Demo:** `UnicodeLookup` tool at the CT website shows the UNICODE of any character you input.

UTF-8

UTF-8 is a particularly efficient and widely used encoding format for Unicode. UTF-8 encodes each Unicode code point using 1 to 4 octets (an octet is 8 bits or a byte in the Unicode standard). Earlier positions in the Unicode character set, which are used more often, require fewer bytes. UTF-8 is US-ASCII compatible because the first 128 Unicode characters are exactly those from US-ASCII and are encoded with the same 8-bit pattern as US-ASCII. Therefore, any valid US-ASCII encoded character is also the same UTF-8 Unicode character.

The UTF-8 encoding format can be explained simply as follows.

- The first 128 code points (`U+0000`–`U+007F`) use one byte, `0` followed by 7-bit US-ASCII code for the target character (0□□□□□□□).

- The next 1920 code points (`U+0080`–`U+07FF`) use two bytes: first byte=`110` followed by 5 bits (110□□□□□), second byte=`10` followed by 6 bits (10□□□□□□). The 11 useful bits are more than enough for the 1920 code points.

- The next 63487 code points (`U+0800`–`U+FFFF`) use three bytes: first byte=`1110` followed by 4 bits (1110□□□□), next two bytes each in the form `10` followed by 6 bits (10□□□□□□).

- The remaining code points (`U+10000`–`U+1FFFFF`) use four bytes: first byte=`11110` followed by 3 bits (11110□□□), the next three bytes each in the form `10` followed by 6 bits (10□□□□□□).

Note with UTF-8, the first byte of a multibyte representation specifies the number of bytes by the count of leading `1` bits. Nonleading bytes all start with `10` to distinguish them from leading bytes.

With UTF-8, US-ASCII coded documents are automatically Unicode texts. UTF-8 is now recommended for all webpages, email messages, programming languages, operating systems, and applications.

Reuse of Bit Patterns

You must have figured out by now that a given bit pattern may represent a binary number or a character. For example, the bit pattern

```
01000001
```

represents 65 or the character `'A'`. The question is, how do we know which one? Again, the answer is "context."

> **CT: DATA CONTEXT**
>
> *The same bit pattern can be interpreted differently depending on the context where it is used.*

We must provide a context to treat any given bit pattern to indicate if it is a number, a character, or something else. The context can be deduced from where the pattern is used or by explicitly provided indications. For example, in evaluating the expression `x > 0`, we know the value of `x` needs to be interpreted as a number. A programming language usually supplies a way to declare the *data type* of a variable. The type informs a program how to interpret the data representation associated with any given variable.

2.7 Editing Text

One of the most basic tasks on a computer is text editing. You would use a text editor to input text, make changes (add, delete, revise text), search for character strings, save the result, and perform other functions. A text file contains *plain text*, meaning that from start to finish,it contains just a string of characters. Data and program source code files on a computer are usually in plain text. Text editors help you produce such plain text files.

Default text editors are usually made available by your operating system, for example Notepad/Wordpad for Microsoft Windows, TextEdit for Mac System X, and vi and gedit for Linux. Text editors that are free to download include vim/gvim (Linux, Windows and Mac), emacs (Linux, Windows and Mac), TextWrangler (Mac) and notepad++ (Windows).

Find a good text editor, learn it well, and you are well on your way to writing and editing computer codes efficiently. The CT website has materials to help you learn Vi/Vim and Emacs.

Word processing programs such as MS Word™, LibreOffice, LaTeX, and Adobe Acrobat™, are different from text editors. These help you create reports, presentations, and other well formatted and styled documents mostly for human consumption. Document files produced by word processors contain formatting and display code so they can be presented nicely for viewing or printing.

Input Methods

A keyboard allows us to type data into a computer. When a key is pressed or released, an event is generated that is detected by the operating system. An operating system has an internal keycode table that assigns a different numerical keycode to each key on the keyboard.

Here are some example keycodes from Linux.

key `ESC`	1	key `TAB`	15
key `A`	30	key `S`	31
key `D`	32	key `L Shift`	42
key `F1`	59	key `UP`	103

A keycode can then lead to a character encoding. Some keycodes, such as those for function keys or arrow keys, correspond to no character code.

The default character encoding is determined by the computer's locale setting. The locale determines the region, language, time, currency, and other such features used by a computer. For example, the locale can be set to "`en_US.UTF-8`" for a computer in the USA.

Using a Western keyboard, we can easily enter US-ASCII characters. For other characters, we need to add language support as well as *input methods* for the desired languages. Having set up multiple input methods, you can then switch from one input method to another at will to enter data from the keyboard.

Figure 2.11 shows three available input methods: English (standard keyboard), Chinese Pinyin, and English (Dvorak keyboard). Usually, CNTL+SPACE (pressing SPACE while holding down the CONTROL key) will switch to the next available input method.

As an example, Figure 2.12 shows the a keyboard layout for entering Traditional Chinese characters using phonetics.

On a touch screen, you can use a specialized on-screen keyboard to enter characters in different languages. Better yet, with voice recognition you can just say the words to create text input. Figure 2.13 shows the Chinese Pinyin keyboard on an Android phone.

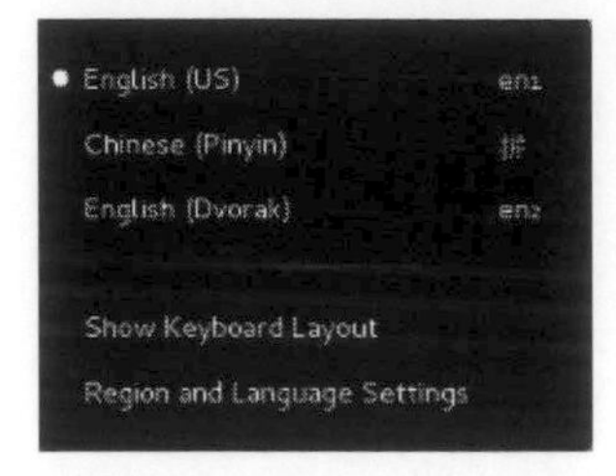

FIGURE 2.11 Input Methods

FIGURE 2.12 Keyboard Layout for Chinese Phonetic Symbols

Modern applications, including Web browsers, email clients, text editors, and document processors support Unicode text, provided that the necessary language support and display fonts have been installed on your system.

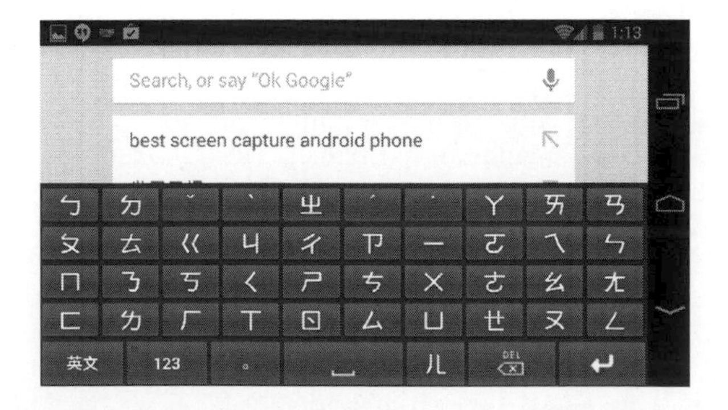

FIGURE 2.13 Chinese Phonetic Input on Android Phone

2.8 Data Output

Data encoded in binary allow fast processing and storage on the computer. Yet, we still need to output data for human consumption. This means displaying numbers, characters, and images, playing audio and video, and so on.

In the early days, operators used card punches to punch holes on stiff paper cards (Figure 2.14) as input to *batch-processing computers* housed in climate-controlled machine rooms. And the computed results were output to line printers. Early line printers were washing-machine sized objects that impact print text, one line at a time, usually on continuous-feed form sheets. Turn-around time was measured in hours and even days.

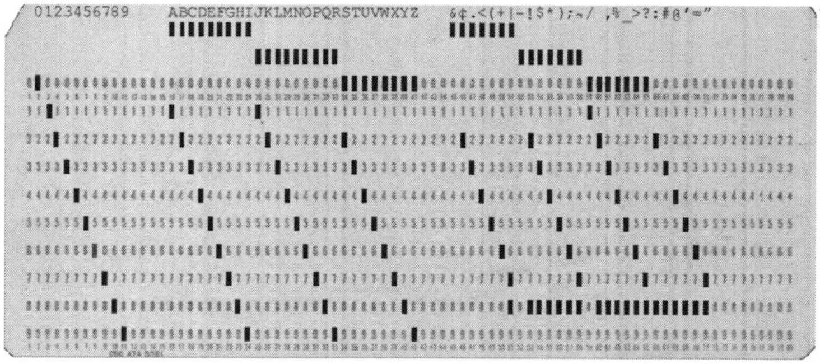

FIGURE 2.14 A Punch Card

Later in history, CRT (cathode ray tube) monitors, connected to any given computer, provided concurrent access (input and output) by multiple users to *time-sharing systems.* Users enjoyed immediate interactions with computers and productivity increased tremendously. CRT monitors are monochrome (one color) and display characters (US-ASCII) with hardware-defined shapes.

Computers today have full-color, high-resolution LCD/LED graphics displays supported by dedicated GPU (Graphics Processor Unit) hardware. The pixel rendering of characters is controlled by *font files.* A font is a particular style or design of the appearance of the characters. For example, *Serif* fonts add little lines at the end of strokes. *Sans Serif* fonts are without such added decorations. Well-known font families include *Times New Roman*, *Helvetica*, and *Courier.* As an example, Figure 2.15 shows a particular Courier font design.

Within a font family, you can also call for a variant style, such as *italic* and **boldface**. Normally, fonts are proprietary, requiring permission/fee for their use. But, free fonts in the public domain do exist. Many fonts have been designed before the age of computers or even typewriters. Typefaces for computers are stored in font files that allow easy size scaling. Your computer comes with a collection of fonts. Additional fonts can be downloaded and installed. Applications usually allow you to pick from available fonts to customize the display of characters. Figure 2.16 shows a set of fonts for Chinese charac-

abcdefghi
jklmnopqr
stuvwxyz
ABCDEFGHI
JKLMNOPQR
STUVWXYZ
123456789

FIGURE 2.15 A Courier Font

	en	ja	ko	zh-cn	zh-hk	zh-tw
sans serif	直	直	直	直	直	直
serif	直	直	直	直	直	直
monospace	直	直	直	直	直	直

	en	ja	ko	zh-cn	zh-hk	zh-tw
sans serif	骨	骨	骨	骨	骨	骨
serif	骨	骨	骨	骨	骨	骨
monospace	骨	骨	骨	骨	骨	骨

AR PL ShanHeiSun Uni	直骨
AR PL ZenKai Uni	直骨
Sazanami Gothic	直骨
Sazanami Mincho	直骨
UnBatang	直骨
UnDotum	直骨

FIGURE 2.16 Chinese Fonts

ters. Different *typefaces* or *fonts* can be made available to support displaying characters in any given language.

The ability to output characters makes displaying numbers a breeze. Numbers represented in binary are converted to decimal (base 10) and the digits (0–9) are displayed as characters for humans to read.

> **CT: Deliver the Message**
>
> *Information must be conveyed in a form that can be understood and processed by the receiver.*

This is so true for human-to-human, human–animal, and human–computer communication, as professor Temple Grandin, famous for her contributions to communications with autistic children and animals alike, would surely tell you.

Exercises

2.1. What is the difference between digital and analog signals? Please explain.

2.2. What is a word? How many bytes in a word? Explain clearly.

2.3. What is a bit pattern? Let n be the number of bits. What is the number of different bit patterns for $n = 3$, $n = 4$, $n = 8$, or $n = 32$?

2.4. What is the decimal value of the binary number `101`? `10101101`?

2.5. Write down the octal and hex representations for the decimal number 911. Show your calculations, as well as the final answers.

2.6. Find the two's complement binary number for -103. Show your calculations, and verify that when it is added to 103, you get zero. Extra credit: Prove that the two's complement algorithm always produces the negative of any m.

2.7. What is an equivalence relation?

2.8. Compute 13^9 mod 16. Hint: The result is between 0 and 15.

2.9. What the difference between text editing and word processing?

2.10. What is UNICODE? UTF-8? What's the difference?

2.11. What is a keycode? Who defines keycodes?

2.12. **Computize**: What does the word "actual" mean in Spanish? In English? What CT principle does it demonstrate?

2.13. **Computize**: The byte `01000010` represents what number? Or what ASCII character? What CT principle does it demonstrate?

2.14. **Computize**: What bit pattern representation is used for `303` as a telephone area code or as the number three hundred and three? How do we tell them apart?

2.15. **Group discussion topic**: *Mary had a little lamb. Context and semantics.*

2.16. In-class group activity: Invite two teams of four students, team A and B, to the front of the class. Each team member has large letters *0,1,...,9,A,B,C,D,E,F*, each letter on a separate sheet of paper. The CT website has ready-to-print letters. Team A is the binary team and B the octal team, say. The class will shout out a number, in a given range, for both teams to form the correct display. The faster team wins and gets to sit down. The winning team members can pick others in class to replace them. The instructor will keep it fun, encouraging, and interesting.

Chapter 3

True or False

Digital computers are logic machines. They use bits to store information. But a bit is nothing but a switch with two states, on and off. A bit can represent the two binary digits `1` and `0`. And a bit can represent the two truth values *true* and *false*.

Computer hardware consists of circuits built with logic gates. Boolean algebra deals with computations on truth values. Logic conditions and implications are used in software to control program execution.

We shall explore the close relationship between logic and computers.

3.1 Digital Electronic Circuits

Modern computers process digital signals represented by the presence (1) or absence (0) of an electric current or voltage. Digital electronic circuits process input signals and produce output signals that represent specific computational results. CPU registers hold input signals as well as store output signals.

Treating 1 as *true* and 0 as *false*, the basic logical operations on the input *truth values*, `A` and `B`, are as follows.

- `(A AND B)` is true only when `A` and `B` are both true.
- `(A OR B)` is true if at least one of `A` and `B` is true.
- `(NOT A)` is true if `A` is false and is false if `A` is true.
- `(A XOR B)` is true if only one of `A` and `B` is true.
- `(A NAND B)` is false only when both `A` and `B` are true.
- `(A NOR B)` is false if at least one of `A` and `B` is true.
- `(A XNOR B)` is false if only one of `A` and `B` is true.

We see seven logic operators listed. The complete behavior of each logic operator can be described by a *truth table* specifying the output for all possible input values.

Table 3.1 shows the truth tables for `AND`, `OR`, and `XOR`. Try it yourself and write down the truth tables for the other logical operators.

TABLE 3.1 Truth Tables

A	B	A AND B	A OR B	A XOR B
0	0	0	0	0
0	1	0	1	1
1	0	0	1	1
1	1	1	1	0

CT: NOTICE THE LOGIC

Be careful about the precise meaning of "and", "or", "not", and "nor". Their meaning as logic operators may not be the same as in day-to-day usage. Guard against their imprecise use when receiving communication from others. Better be careful than sorry.

For example, the menu in a restaurant may state "mashed potatoes, French fries, or steamed broccoli." It usually means you can choose only one. Whereas, the job requirement "a bachelor or master's degree" means either or both.

Logic Gates

A *gate* is a basic building block in integrated circuits constructed with CMOS (complementary metal–oxide–semiconductor) or similar technologies. A *logic gate* implements one of the basic logic operations, as we have just described. A logic gate takes one or two inputs and produces a single output. The seven basic gates are as follows. Each gate, except the `NOT` gate, takes two inputs.

- `AND` gate—Outputs 1 only if both inputs are 1. Otherwise outputs 0.
- `OR` gate—Outputs 0 only if both inputs are 0. Otherwise outputs 1.
- `NOT` gate—Outputs the opposite of the input.
- `XOR` (exclusive `OR`) gate—Outputs 1 only if the two inputs are different. Otherwise outputs 0.
- `NAND`, `NOR`, or `XNOR` gate—Outputs the opposite of the corresponding `AND`, `OR`, or `XOR` gate.

Standardized graphical symbols can be used to design simple digital circuits. Figure 3.1 shows the traditional distinctive shape symbols. Note that a

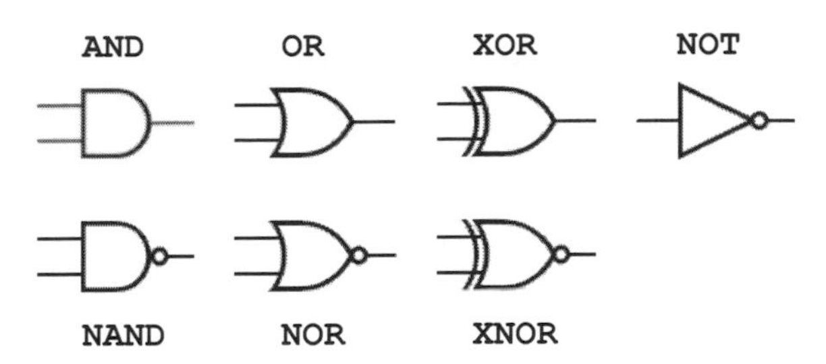

FIGURE 3.1 Distinctive Gate Symbols

little circle at the tip of a gate indicates an inversion, turning a 1 to 0 and a 0 to 1. Figure 3.2 shows the newer rectangular shape symbols.

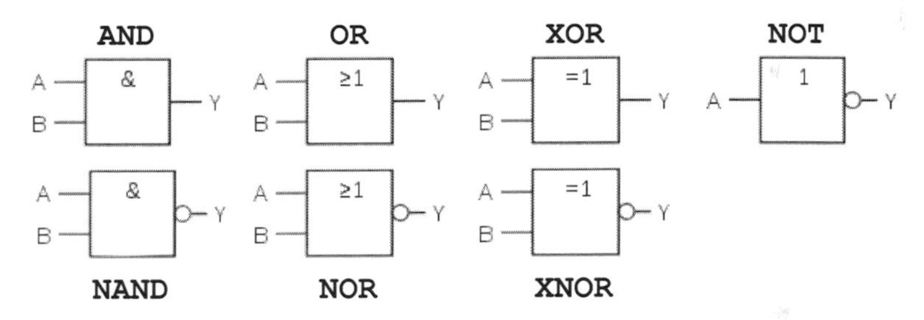

FIGURE 3.2 Rectangular Gate Symbols

More complicated digital circuits are built by wiring together logic gates. Pulse signals, typically produced by a crystal oscillator, provide a processing rhythm. The next processing step won't start until a new clock pulse arrives. The pulse interval is needed for all digital signal transitions to complete and the circuit components to settle into their new states. Typical clock rates for modern CPUs are in the GHz range.

Adders

More complicated digital circuits can be built by wiring together logic gates. On a CPU, arithmetic operations on binary numbers are the most basic. An *adder* circuit provides the ability to add binary numbers, which in turn can be used repeatedly to perform multiplication of binary numbers. As an example, let's look at how an adder can be built using logic gates.

But first, we will build a *half adder*. A *half adder* (Figure 3.3) takes two input bits `A` and `B` and outputs a sum bit `S` and a carry bit `C`. In Figure 3.3, we used an `XOR` gate to produce the sum bit and an `AND` gate to compute the carry bit. Table 3.2 lists the truth values for the half adder.

You can go to the CT website and play with an interactive tool to exper-

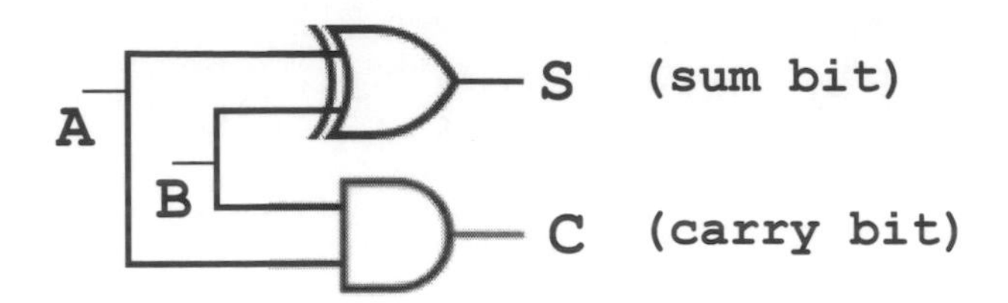

FIGURE 3.3 A Half Adder

TABLE 3.2 Truth Table for Half Adder

A	B	S	C
0	0	0	0
0	1	1	0
1	0	1	0
1	1	0	1

iment with logic gates (**Demo:** `GateSimulator`). Why not try to build your own half adder using the gate simulator?

A *full adder* takes three input bits, A, B, and C_{in} (carry in) and produces a sum bit, S_{out}, and a carry bit, C_{out}. It can be built with two half adders and an extra `OR` gate, as shown in Figure 3.4. Table 3.3 shows the truth table for the full adder.

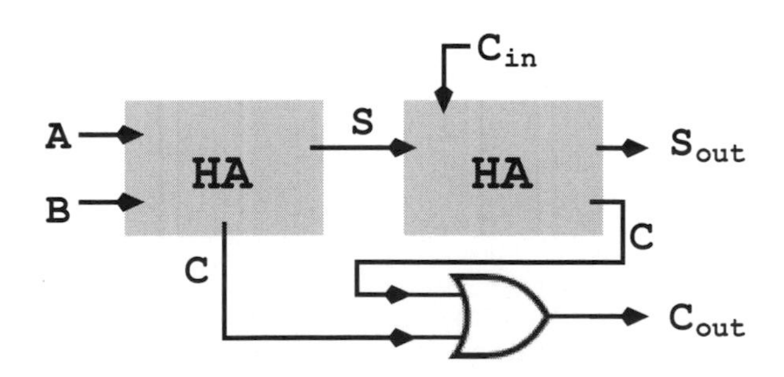

FIGURE 3.4 A Full Adder

Building simple components first then combining them into more complicated components is a solution method that often works well.

TABLE 3.3 The Truth Table for a Full Adder

A	B	C_{in}	S_{out}	C_{out}	A	B	C_{in}	S_{out}	C_{out}
0	0	0	0	0	0	0	1	1	0
0	1	0	1	0	0	1	1	0	1
1	0	0	1	0	1	0	1	0	1
1	1	0	0	1	1	1	1	1	1

CT: BOTTOM UP

Consider the bottom-up approach for problem solving. It can be effective in many situations.

We first built a half adder. Then we combine two half adders to build a full adder. Then we can combine a number of full adders to solve the more complicated problem of adding two n-bit numbers.

To add two n-bit binary numbers (unsigned or two's complement), we can combine n full adders, each taking its Cin from the $Cout$ of the adder to its right. In Figure 3.5 we see how this works for input `X` ($X_n X_{n-1} ... X_0$) and `Y` ($Y_n Y_{n-1} ... Y_0$) producing a result `S` ($S_n S_{n-1} ... S_0$).

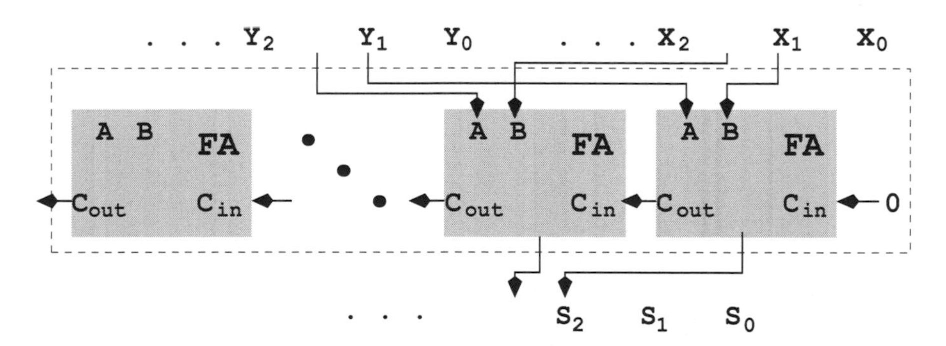

FIGURE 3.5 A Ripple Adder

Because the carry bit may need to propagate from the rightmost adder all the way to the leftmost, a delay proportional to n is needed until the last output bit settles down. Such an adder is known as a *ripple adder*. More complicated adders can reduce the delay and make binary addition faster, especially if addition is performed as a subprocess of multiplication.

Computer hardware design is a complicated engineering activity. Software tools are used to help automate various phases of the design, testing, synthesis, and fabrication of integrated circuits. Better hardware leads to faster software, which leads to even better hardware, a wonderful, virtuous cycle, indeed.

CT: Create a Virtuous Cycle

Take advantage of new knowledge, new tools, new circumstances and apply them everywhere possible. Don't miss the chance to create a positive feedback loop!

For example, repeated iteration of a process has led to the invention of the *polymerase chain reaction* (PCR), a technique in molecular biology to generate thousands to millions of copies of a particular DNA sequence. Developed by Dr. Kary Mullis in 1983, PCR is now indispensable in medical and biological research and applications, including DNA testing and genetic fingerprinting. The impact of automated PCR is huge and far-reaching. Mullis was awarded the 1993 Nobel Prize in Chemistry for his part in the invention of PCR.

In recounting his invention, Dr. Mullis wrote in his book *Dancing Naked in the Mind Field*:

> *I knew computer programming, and from that I understood the power of a reiterative mathematical procedure. That's where you apply some process to a starting number to obtain a new number, and then you apply the same process to the new number, and so on. If the process is multiplication by two, then the result of many cycles is an exponential increase in the value of the original number: 2 becomes 4 becomes 8 becomes 16 becomes 32 and so on.*
>
> *If I could arrange for a short synthetic piece of DNA to find a particular sequence and then start a process whereby that sequence would reproduce itself over and over, then I would be close to solving my problem.*

At the time of the invention, the "polymerase" and other related DNA duplication techniques were already known. It was the "chain reaction" part that was missing. Well, we have Dr. Mullis and his computational thinking to thank for the invention. And what a significant invention! *The New York Times* described it as "highly original and significant, virtually dividing biology into the two epochs of before P.C.R. and after P.C.R."

Still need more convincing? Just ask the Innocence Project or any guiltless person freed from jail due to genetic fingerprinting.

3.2 Boolean Algebra

The word *algebra* comes from Arabic *al-jebr*, meaning "reunion of broken parts." Elementary algebra, the kind we learn in middle school, deals with real numbers and symbols. The symbols stand for variables and unspecified numbers. Boolean algebra, introduced by George Boole in 1854, deals with truth values, *true* and *false* or 1 and 0, instead of numbers. Variables in Boolean algebra may take on either of the two values.

Boolean algebra has the following *basic* operations and operators.

- *Conjunction*—Denoted `A ∧ B`, `A AND B`, `A & B`, or `A • B`; the value of `A AND B` is *true* only if both `A` and `B` are *true*
- *Disjunction*—Denoted `A ∨ B`, `A OR B`, `A || B`, or `A + B`; the value of `A OR B` is *true* if at least one of `A` and `B` is *true*
- *Negation*—Denoted `¬A`, `NOT A`, `!A`, or $\overline{\mathtt{A}}$; the value of `NOT A` is *true* if `A` is *false* and is *false* otherwise

The preceding varied notations remind us of CT: Meaning of Symbols, Section 2.2.

Boolean algebra deals with expressions involving these operators, their properties, and manipulations. As such, it is very useful in the study and design of digital circuits.

3.2.1 Expressions and Laws

Here, in our introduction to Boolean algebra, we will use the values `0` and `1`, and the operators •, +, and ‾.

Let a, b, and c be Boolean variables. The following *laws* hold in Boolean algebra.

- Simplification laws: $a \bullet a = a,\ a + a = a,\ \overline{(\overline{a})} = a,\ 0 + a = a,\ 0 \bullet a = 0,\ 1 + a = 1,\ 1 \bullet a = a$
- Communicative laws: $a \bullet b = b \bullet a,\ a + b = b + a$
- Associative laws: $a \bullet (b \bullet c) = (a \bullet b) \bullet c,\ a + (b + c) = (a + b) + c$
- Distributive laws: $a \bullet (b + c) = (a \bullet b) + (a \bullet c),\ a + (b \bullet c) = (a + b) \bullet (a + c)$
- Absorption laws: $a \bullet (a + b) = a,\ a + (a \bullet b) = a$ (The variable b is absorbed as if it is not there.)
- Negation laws: $a \bullet \overline{a} = 0,\ a + \overline{a} = 1$
- De Morgan's laws: $\overline{a \bullet b} = \overline{a} + \overline{b},\ \overline{a + b} = \overline{a} \bullet \overline{b}$

The three logical operations `AND`, `OR`, and `NOT` are basic in that they can be used to produce any truth table with up to 2 inputs and one output. For example, a XOR b $= (a + b) \bullet \overline{(a \bullet b)}$.

3.2.2 Universal Gates

A logic gate is universal if it alone can be used to implement any Boolean function. The NAND and NOR gates are universal. Because NAND and NOR gates are economical and easier to fabricate, they are the basic gates used in all IC (integrated circuit) digital logic families. In practice, an AND (OR) gate is usually implemented as a NAND (NOR) gate followed by an inverter.

Let's prove that NAND is universal by showing how it is used to build AND, OR, and NOT gates. Boolean algebra come in handy.

- Building NOT gate—$\overline{(a \bullet a)} = \overline{a}$, $\overline{(a \bullet 1)} = \overline{a}$
- Building AND gate—$\overline{(\overline{(a \bullet b)})} = a \bullet b$
- Building OR gate—$\overline{(\overline{a} \bullet \overline{b})} = a + b$

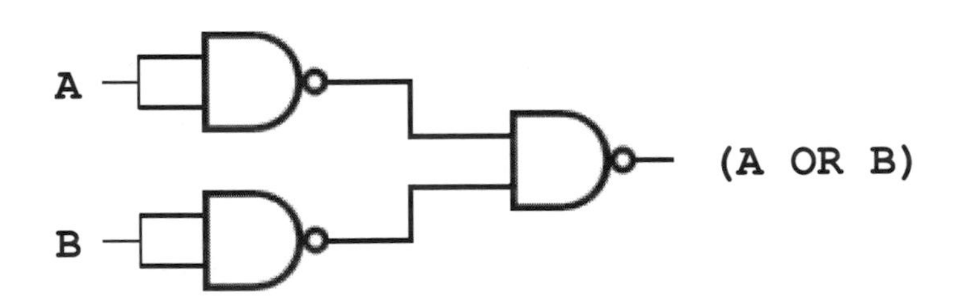

FIGURE 3.6 Implementing OR with NAND Gates

It would be fun to use the GateSimulator tool on the CT website to build AND, OR, and NOT gates using either NAND or NOR gates exclusively (**Demo:** NandGate). See Figure 3.6 for implementing an OR with NAND gates.

3.3 Decision Making

When drawing a flowchart (Section 1.7) or specifying an algorithm (Section 1.8), we often need to have test conditions. Depending on the yes/no answer of a test, a procedure may take a different path through its steps.

In logic and in programming, a function that produces a result that is either *true* or *false* is known as a *predicate*. Programming languages usually provide *relational operators* as predefined predicates. Table 3.4 lists relational operators in JavaScript that compare numerical quantities and produce a truth value.

Usually, a bit pattern with all 0s is treated as *false*, and any other value is

TABLE 3.4 JavaScript Relational Operators

Operator	Meaning
`==`	Equal to
`!=`	Not equal to
`<`	Less than
`>`	Greater than
`<=`	Less than or equal to
`>=`	Greater than or equal to

treated as *true*. This makes sense because 0 is *false* and anything that is not 0 is *true*. An immediate result of this convention is that any function that returns a value can be treated as a predicate. Programming languages also provide *logical operators* to perform Boolean operations on truth values. Table 3.5 lists

TABLE 3.5 JavaScript Logical Operators

<table>
<tr><th>Operator</th><th>Meaning</th></tr>
<tr><td><code>&&</code></td><td><code>AND</code></td></tr>
<tr><td><code>||</code></td><td><code>OR</code></td></tr>
<tr><td><code>!</code></td><td><code>NOT</code></td></tr>
</table>

JavaScript logical operators. Like most other languages, JavaScript adopted relational and logical operators from C/C++.

As an example, let's use the preceding notations to define a predicate, **`rhNormal`**, which takes an input relative humidity reading and determines if it is in the normal comfortable range (between 50% to 60%) for people.

Algorithm **`rhNormal`**:
Input: Integer percentage **`rh`**
Output: Returns `0` (*false*) or nonzero (*true*)

1. If `(rh > 60 || rh < 50)`, then return `0`
2. Return `1`

Alternatively, we can use the following.

Algorithm **`rhNormal`**:
Input: Integer percentage **`rh`**
Output: Returns `0` (false) or nonzero (*true*)

1. If `(rh >= 50 && rh <= 60)`, then return `1`
2. Return `0`

CT: Logic Checks

Computerized predicate checking can help automate many tasks, large and small.

Automatic control often means keeping values of certain parameters within allowable ranges. A humidity control system may call `rhNormal` periodically and decide to start/stop the increasing or decreasing the humidity. Such computer controls are commonplace. You'll find them in automobile cruise control systems, antilock brake systems, GPS navigation systems, autopilot for airplanes, and so on.

3.3.1 Conditions and Implications

As you can imagine, when devising an algorithm, making the right decisions on which next step to take is critical. Typically, we use

`if` *predicate*, `then` $action_1$, `else` $action_2$

to indicate such decisions. If the *predicate* evaluates to *true*, $action_1$ is taken. Otherwise, $action_2$ is taken. The `else` part is usually optional in the notation.

Correctness of an algorithm depends on using the right *implications*. An implication is a logical statement commonly given in these forms.

- p implies q, or $p \implies q$
- if p, then q
- q if p,

where p is a *premise* and q is a *conclusion*.

Let's look at an algorithm that compares two input numbers, `x` and `y`, and

1. Returns a positive number if `x` is larger than `y`;
2. Returns a negative number if `x` is smaller than `y`; and
3. Returns zero if `x` is equal to `y`.

Algorithm `numberCompare`:
Input: Number `x`, number `y`
Output: Returns `1`, `0`, or `-1`

1. If `x > y`, then return `1`
2. If `x < y`, then return `-1`

3. Return `0`

In algorithm `numberCompare`, note the implications

"control flow reaching Step 2" $\Longrightarrow$ "`x <= y`"

"control flow reaching Step 3" $\Longrightarrow$ "`x == y`"

Now think about why `numberCompare` can be implemented simply as "`return x - y`.".

Given the implication $p \Longrightarrow q$ (p being *true* causes q to be *true*), then the following statements are true.

- p is a *sufficient condition* for q, namely, p being *true* guarantees that q is *true*.
- q is a *necessary condition* for p, namely q must be *true* for p to be *true*. Also, if q is *false*, then p is *false* as well. Thus, the implication $p \Longrightarrow q$ is logically the same as the implication $\overline{q} \Longrightarrow \overline{p}$.
- If p is *false*, the implication says nothing about q.
- If q is *true*, the implication says nothing about p.

For example, the implication, "If x is a woman then x is a person," certainly does not mean, "If x is a person then x is a woman." Nonetheless, if x is not a person, then x can not be a woman.

Similarly, "It is a river $\Longrightarrow$ water flows in it," does not mean if water flows in it, then it is a river. In fact, it could be a water hose or a drain pipe. But, if water does not flow in it, then it is not a river.

And, "If n is a multiple of 8 $\Longrightarrow$ n is an even number," does not mean if n is even, then it is divisible by 8. And, "If a person is over 30 years old, then the person is an adult," does not mean an adult is over 30.

Finally, "A good computer programmer thinks logically," does not mean that anyone who thinks logically is a good programmer. The person must have other training as well. Yet, it is definitely the case that without logical thinking one cannot be a good programmer.

To summarize, a sufficient condition may not be necessary and a necessary condition may not be sufficient.

However, if we have both implications $p \Longrightarrow q$ and $q \Longrightarrow p$, then q is a *necessary and sufficient* condition for p. Likewise, p is a *necessary and sufficient* condition for q. Alternatively, we say p if and only if q or simply $p \Longleftrightarrow q$. In such a case, p and q are both *true* or both *false*. For example, a person may vote in a United States election if and only if the person is a United States citizen, at least 18 years old, and not a convicted felon.

CT: Follow the Logic

Think all this logical stuff is simple and straightforward? Think again!

Don't hesitate to study the materials over again. Make logic your own natural mental tool, and it will help immensely in whatever you do. However, once logic is natural to you, don't assume others are the same. In fact, it is a good bet to assume otherwise. Your being logical certainly does not imply that everyone else is. Because we need to work with others to achieve many tasks, guarding against falling victim to less than logical thinking on the part of others would be wise indeed.

3.4 Logic Applied to Bits

We have already seen basic operations on data, including arithmetic, logical, and relational operations. Such operations deal with whole bytes or words.

But, because data are represented by bit patterns, we sometimes also want to operate on individual bits. Programming languages provide *bitwise* operators for this very purpose.

For example, the *bitwise operator* `&`, applies logical `AND` at the bit level. Table 3.6 shows `b & m`, the bitwise AND operation on two bytes `b` and `m`.

TABLE 3.6 Bitwise AND

b	1	1	0	0	1	1	0	1
m	0	0	0	0	1	1	1	1
b & m	0	0	0	0	1	1	0	1

Each bit in the result is computed as the logical AND of the corresponding pair of bits in `b` and `m`. Thus, each result bit is computed from the two bits above it in the same column.

Often, we need to focus on a certain subset of bits in a bit pattern. For any bit pattern `b`, we can use `b & m` to extract arbitrary desired parts of `b`. Here, `m` is known as a mask. A `1` bit in the mask selects that corresponding bit we want from `b`, whereas a `0` bit in the mask blocks an unwanted bit from `b`. To visualize the masking, think of bits in `b` as contestants in a game. Judges give a `1` to advance a contestant and a `0` to reject another. Thus, the 1-bit part of a mask is *transparent* and the 0-bit part *opaque*.

For example, to test if any integer `n` is odd, we can use the predicate `(n & 1)`. The bit pattern for the mask `1` has all zeros except the rightmost bit.

TABLE 3.7 JavaScript Bitwise Logical Operators

Operator	Meaning	Operator	Meaning
&	`Bitwise AND`	|	`Bitwise OR`
~	`Bitwise NOT`	^	`Bitwise XOR`

Table 3.7 lists all bitwise logical operators in JavaScript. Using bitwise NOT, we can implement the two's complement algorithm (Section 2.3) this way:

Algorithm `2sComplement`:
Input: Integer `a`
Output: Returns `-a`

1. Set `a = ~a`
2. Return `a + 1`

Now, let's see how bitwise XOR works. Table 3.8 shows an example of the bitwise XOR demonstrating for any `a`

`a ^ a = 0.`

TABLE 3.8 Bitwise XOR

a	1	1	0	0	1	1	0	1
a	1	1	0	0	1	1	0	1
a ^ a	0	0	0	0	0	0	0	0

Of course, we also know that for any `a`

`a ^ 0 = a.`

Let `r` and `s` be any pair of bit patterns. We can interchange their values without using a third temporary variable. Let's consider the sequence of operations:

1. Set `r = r ^ s`
2. Set `s = r ^ s`
3. Set `r = r ^ s`

Would you believe that these three steps swap the values of `r` and `s`? That is, after the three steps, `s` gets the original value of `r`, and `r` gets the original value of `s`.

To show this, our understanding of XOR comes in handy. Let `r0` be the original `r` and `s0` be the original `s`. Taking the value of `r` from Step 1 and substituting it into Step 2, we have

```
s = r0 ^ s0 ^ s0 = r0 ^ 0 = r0.
```

Thus, the value of `s` is `r0` after Step 2.

Now, substituting the value of `r` (from Step 1) and the value of `s` (from Step 2) into Step 3, we get

```
r = r0 ^ s0 ^ r0 = r0 ^ r0 ^ s0 = 0 ^ s0 = s0.
```

Alas, we have swapped the values of `r` and `s` without using a third variable. This example is cute in showing bitwise logical operations. But, the simple swap:

Set `temp = r`
Set `r = s`
Set `s = temp`

is still the recommended approach.

We can also take a bit pattern and shift it left or right a number of bits. For example,

Set `x = a >> 3` (`x` is `a` shifted right 3 bits)
Set `x = a << 4` (`x` is `a` shifted left 4 bits)

Bits shifted out (off the end) are discarded. New bits shifted in are zero.

If `a` is an unsigned number, then we have

`a << 1` (gives `a` × 2)
`a >> 1` (gives quotient of `a / 2`).

As another application of bitwise operations, let's look at the following algorithm, which displays the binary representation of any input byte `b`.

Algorithm `byteDisplay`:
Input: Byte `b`
Output: Displays the bits of `b`

1. Set `m = 1 << 7`
2. If `m == 0`, then display NEWLINE and terminate
3. If `(b & m)`, then display `1` else display `0`
4. Set `m = m >> 1`

5. Go to Step 2

Trace the algorithm to see how it works. Convince yourself that the condition `(m == 0)` can be replaced by `(! m)`.

> **CT: Combine Basic Components**
>
> *Complicated systems often consist of simple elementary parts that interact following simple rules.*

We see sophisticated computer systems, those that sent space probes to Mars or men to the moon and back, are composed of bits and basic logic operations, implemented by NAND gates.

Similarly, in biology, atoms form molecules that form cells in plants and animals. Computers and living things are not that different after all.

3.5 Logic and Iteration

Often, the power of a computer program rests in its ability to perform *iterations* or to repeat the same steps over and over again. A set of steps for repeated execution is called a *loop*. Central to any loop is *loop control*, the conditions and actions that determine how to start, perform, and end the iterations.

3.5.1 The `while` Loop

Programming languages usually provide constructs to support writing loops. Let's look at a construct for the `while` loop first. A `while` loop usually takes the form

`while (` *condition* `)` (loop control)
`{` *steps to perform* `}` (loop body).

A `while` loop basically says, "While the *condition* holds true, repeatedly perform the loop body." The precise semantics of the `while` loop is given by the flowchart in Figure 3.7.

The `while` loop control is a *continuation test*. It is evaluated before each execution of the loop body. If the test returns true, the loop body is performed once. The test is evaluated again before the next iteration of the loop body. If the test returns false, the while loop is finished.

A *prime number* n is an integer greater than 1 that has no divisors other than 1 and n. Let's write a predicate `isPrime` that determines if a given input integer is a prime number.

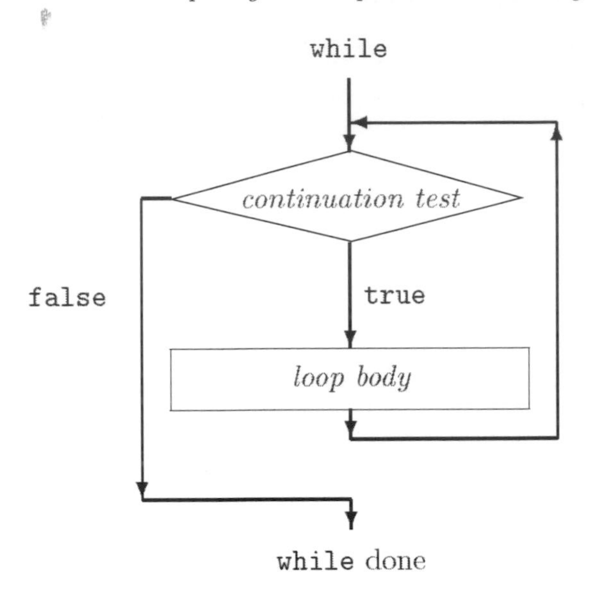

FIGURE 3.7 The `while` Loop

Algorithm `isPrime`:
Input: Integer `n`
Output: Returns `0` (false) or `n` (true)

1. If `n < 2`, then return `0`
2. If `n < 4`, then return `n`
3. If (`remainder(n,2)==0`), then return `0`
4. Set `d = 3`
5. `while ( d < n )`
   `{` If (`remainder(n,d)==0`), then return `0` else set `d = d+2 }`
6. Return `n`

The algorithm is given to illustrate the `while` loop. It is far from an efficient prime testing algorithm. For one thing, the `while` loop can stop much earlier (`d <= sqrt(n)`) by using the square root of `n`.

Trace the algorithm to see how it works. Make sure you try the values `0` through `3` for `n`.

3.5.2 The `for` Loop

A `for` loop takes the form

for (*initialization*; *continuation test*; *increment*) (loop control)
{ *steps to perform* } (loop body).

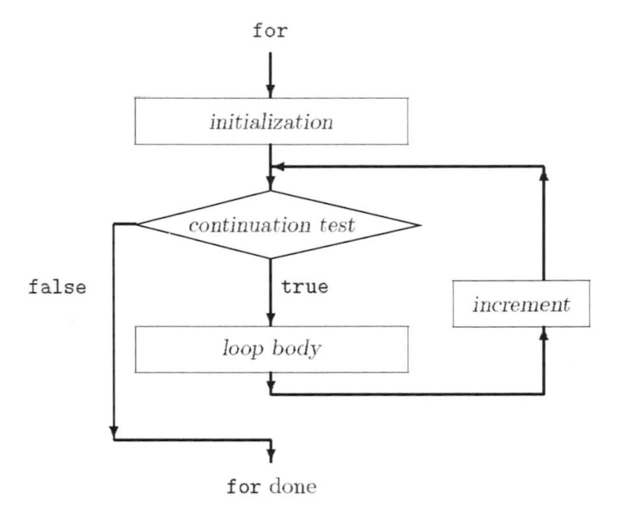

FIGURE 3.8 The `for` Loop

Note that

- The *initialization* part of loop control is evaluated only once in preparation for performing the loop.
- The *continuation test* is evaluated once before each iteration of the loop body. If the test produces false, the loop is finished.
- The *increment* part is evaluated once after each iteration of the loop body.

Thus, after each iteration, we evaluate *increment* then evaluate *continuation test* to see if we need to do another iteration.

An immediate improvement to the preceding `isPrime` predicate can be made by using a list of small primes, say, all primes from 2 to 997:

```
primes = [2,3,5,7,11,13,17,19,23,29,31,37,41,43,47,53,
59,61,67,71,73,79,83,89,97,101,103,107,109,113,127,131,
137,139,149,151,157,163,167,173,179,181,191,193,197,199,
211,223,227,229,233,239,241,251,257,263,269,271,277,281,
283,293,307,311,313,317,331,337,347,349,353,359,367,373,
379,383,389,397,401,409,419,421,431,433,439,443,449,457,
461,463,467,479,487,491,499,503,509,521,523,541,547,557,
563,569,571,577,587,593,599,601,607,613,617,619,631,641,
643,647,653,659,661,673,677,683,691,701,709,719,727,733,
```

```
739,743,751,757,761,769,773,787,797,809,811,821,823,827,
829,839,853,857,859,863,877,881,883,887,907,911,919,929,
937,941,947,953,967,971,977,983,991,997]
```

The variable `primes` holds a list of values instead of just one. In programming, the list of values is called an *array*. Elements in an array are accessed by *indexing*. For example,

`primes[0]` is 2
`primes[1]` is 3
`primes[2]` is 5

and so on. The number of elements in `primes` is `primes.length`.

We can decide if a number n is divisible by any number on `primes` with a `for` loop.

Algorithm `smallprimeCheck`:
Input: Positive integer `n >= 2`, array `primes`
Output: Returns `0` (n not prime), `n` (a prime), last number on `primes` (no divisor)

1. If (`primes contains n`), then return `n`
2. If (`n < primes[primes.length-1]`), then return `0`
3. Set `r = sqrt(n)`
4. `for ( i=0; i < primes.length; i=i+1 )`
   `{` Set `d = primes[i];`
   If (`d > r`), then return `n;`
   If (`remainder(n,d)==0`), then return `0 }`
5. Return `d`

Now the slightly improved `isPrime`.

Algorithm `isPrime2`:
Input: Integer `n`, array `primes`
Output: Returns `0` (false) or `n` (true)

1. If `n < 2`, then return `0`
2. Set `d = smallprimeCheck(n, primes)`
3. If (`d == 0`), then return `0`
4. If (`d == n`), then return `n`
5. Set `d = d + 2` and set `r = sqrt(n)`

6. `while ( d <= r )`
   `{` If (`remainder(n,d)==0`), then return `0` else set `d = d+2 }`

7. Return `n`

Step 2 calls `smallprimeCheck`. If the return values is `0`, then `n` is not a prime (Step 3). If the value is `n`, then it is a prime determined by `smallprimeCheck`. Otherwise, we need to try more divisors, odd `d` up to `d <= sqrt(n)`[1].

Instead of explicit indexing to run through elements of an array, some programming languages allow you to use a `foreach` loop.

```
foreach ( x in array )   (loop control)
{ steps to perform }     (loop body),
```

where the variable *x* automatically takes on the value of the next element in the given *array* before each iteration of the loop body.

For a well-written loop, attention must be paid to the following aspects.

- Initialization—Setting up initial values before the loop begins.
- Termination—Using correct logical conditions for continuing or terminating iterations.
- Continuation—At the end of each iteration, updating values for the next iteration
- Efficiency—Keeping work inside the loop body to a minimum to avoid unnecessary repetition

Efficient and well-designed computer programs require expertise in programming as well as expertise in the problem domain. Primality testing is a perfect case in point. Currently the most efficient general-purpose primality testing algorithm is *elliptic curve primality proving* (ECPP), which depends on sophisticated results from the theory of elliptic curves[2].

The program PRIMO implements ECPP. The record as of 2013 is the prime

$$2^{73845} + 14717,$$

which has 22230 decimal digits. The certification of this number was done by Peter Kaiser with PRIMO 4.0.1. The parallel processing took about seven months on 16 Xeon cores.

[1] Please don't try this algorithm on large numbers. It is too slow.

[2] The Miller–Rabin probabilistic primality test is slightly faster but nondeterministic.

CT: Perform Everyday Programming

Plan your important tasks, at home or in the office, as if writing a program. Execute your tasks carefully and precisely, as if running a program.

Careful planning; explicit checking of required conditions; correct start, finish, and sequencing of steps; taking into account all possible scenarios; and being mindful of efficiency and possible reuse of tried-and-true procedures can all help avoid mistakes and insure success.

Once upon a time, the author was taking his beloved mother, who was visiting him from Taiwan, for a tour of Niagara Falls. Driving from Boston MA, he was excited to show mother the sights. After long hours of driving, they arrived on a Saturday afternoon. Sadly, it turned out mother cannot go across to Canada because she had no visa to Canada! All is not lost. Mother and son took full advantage of the US side of the falls. If the author had planned the trip more carefully, as CT would suggest, then perhaps the trip would be a total success.

Using an incorrect procedure, neglecting something important, or failing to follow correct procedure precisely may have grave consequences. On August 16, 1987, Northwest Airlines flight 255 crashed shortly after takeoff at Detroit Metropolitan Wayne County Airport. All 149 passengers, except a 4-year-old child, and all 6 crew were killed.

The cause? The US National Transportation Safety Board determined that the probable cause of the accident was the flight crew's failure to use the taxi checklist to ensure that the flaps and slats were extended for takeoff.

Exercises

3.1. What is the number of different Boolean operations that take two operands and produce one result? Show your reasoning and list the truth tables for each operation.

3.2. What is the difference between a half adder and a full adder? Please explain in detail.

3.3. State and prove the De Morgan's laws in Boolean algebra.

3.4. How would you implement `AND` using only `NAND` gates? Show the circuit diagram.

3.5. Show that `NOR` is a universal gate.

3.6. Use the gate simulator on the CT website to build a full adder.

3.7. What is a predicate? What are relational operators? Logical operators?

3.8. Are the conditions `if ( a == 0 )` and `if (! a)` the same or not? Explain why.

3.9. What is a necessary condition? A sufficient condition? A necessary and sufficient condition? A necessary condition that is not sufficient? A sufficient condition that is not necessary? Use examples to explain.

3.10. For an integer $n > 2$, show that if n has a divisor $1 < d < n$, then it has a divisor $d <= sqrt(n)$.

3.11. Consider the condition `!( a || b )` and the condition `(!a && !b)`. Are they the same or different? Prove it.

3.12. Trace algorithm `byteDisplay` and see how it works.

3.13. Write the pseudo code for a predicate `isMember` taking an integer n and an array arr of integers. The procedure returns -1 if n is not on the array or an index where n is found on the array.

3.14. **Computize**: Describe the bottom up problem solving approach and one or more real-life application that you find.

3.15. **Computize**: Apply logic and create a "birthday party planning" program in plain human language. Remember to anticipate unforeseen situations.

3.16. **Computize**: Find out more about the PCR process in DNA duplication.

3.17. **Group discussion topic**: *Differences and similarities of the meaning of "AND,", "OR," and "NOT" in English vs. in Boolean algebra.*

3.18. **Group discussion topic**: *If I were rich, then I would be happy.*

3.19. **Group discussion topic**: *PCR and computing, in relation to the human genome project.*

Chapter 4

Who Is the Master?

The operating system (OS) is the master program that runs on your computer, be it a smartphone, tablet, laptop, or desktop. The operating system controls all aspects of a computer and makes it functional as well as usable by people (See Figure 1.7 in Chapter 1). The OS brings life to the innate electronic hardware components and orchestrates all activities on a computer. The same hardware under a different operating system is literally a different computer.

4.1 What Is an Operating System?

The operating system is software that controls and serves all things on a computer, including users, applications (apps[1]), display, I/O (input/output) devices, memory, communication network, files, and security. Operating systems are among the most complex objects humans have ever built. Widely used operating systems include

- Microsoft Windows™—Various versions from Microsoft, such as Windows XP, Windows 7, Windows 8, Windows 10.
- Mac OS X™—Various versions from Apple, such as Leopard, Snow Leopard, Lion, Mountain Lion, Mavericks, Yosemite and so on.
- Linux—Various free (open source) and proprietary distributions, such as Ubuntu, Red Hat Enterprise Linux, CentOS, Fedora, openSUSE, and Debian.
- Android™—Various versions from Google for smartphones and tablets, such as Gingerbread, Honeycomb, Ice Cream Sandwich, Jelly Bean, KitKat and Lollipop.
- iOS™—Various versions from Apple for iPhones and iPads, such as iOS 4, OS 5, iOS 6, iOS 7 and iOS 8.
- Chrome OS™—A Linux-based OS for Web thin client portable computers called Chromebooks.

[1] App is short for application, especially on mobile devices.

Of course, the whole purpose for the OS is for the user to conveniently use the computer and run application programs (usually not part of the OS). Name something you wish to do, and there is likely an application for it. Popular applications include document processors, Web browsers, photo/image editors, CD/DVD burners, and email handlers. Others allow you to text/chat online, send/receive instant messages, make phone/video calls, record/edit audio and video, do accounting and bookkeeping, and prepare tax returns. The list goes on.

More often than not, application programs are written targeting particular operating systems. To run such an application under a different OS, significant reprogramming and testing are required. Thus, operating systems distinguish themselves not only by their own features but also by the richness of applications available.

4.2 Operating System Kernel

Typically, an operating system consists of a *kernel* and a set of system programs. The kernel controls hardware and supports fundamental services, such as I/O control, concurrent program execution, memory management, file services, and network interfaces. The kernel provides a set of *system calls* for other programs to access kernel services. The kernel also interfaces to *device drivers*. A device driver is an OS-specific program for controlling a certain hardware device, such as a network interface card (NIC), webcam, printer, scanner, or mouse. Device drivers are installed and uninstalled as devices get connected or removed.

At any time, a computer can be running in one of two modes.

- *User mode*—The computer is executing code in a program that is not part of the kernel. For example, the user is doing word processing, reading email, or surfing the Web. In user mode, the executing program does not have direct access to hardware, cannot perform restricted instructions, and has no access to memory not allocated to it by the OS. Code in user mode can make a system call to obtain kernel services. A system call switches from user mode to kernel mode. A return from system call switches back to user mode.

- *Kernel mode*—The computer is executing OS kernel code, most likely due to a system call. In kernel mode, the executing code has complete and unrestricted access to the underlying hardware, can execute any CPU instruction, and reference any memory address.

4.2.1 System Programs

In addition to the kernel, modern operating systems usually also supply a set of system programs to make the computer usable and functional. A system program is simply an application that is often packaged together with the OS distribution to supply critical functionalities. System programs will usually provide

- Graphical user interface (GUI) support—Graphical display responsive to user actions using the mouse, keyboard, touch pad, or touch screen (Section 4.4).
- Command-line interface (CLI) support—Terminal emulator (tty), Shell, and Shell-level commands (Section 4.7).
- File management support—Access and control of files and folders by users and applications (Section 4.8.2).
- Web browsing support—Web browser, such as Internet Explorer™, Safari™, or Chrome™.
- Network service support—Servers and clients for Internet services (Chapter 5).
- Media handling support—Playing audio, video, and other media files.
- Languages and locale support—Making the OS useful in different geographical locations and languages.
- System security—Antivirus, firewall, and file encryption programs.
- Program development support—Compilers, linker/loader, and text editors.

4.3 Open Source Software

We understand that big projects always require extraordinary time and effort to complete and involve significant investments. Software development is no exception. This is especially true for large and complicated software, such as operating systems.

In order to protect the investment, finished products are usually heavily protected by patents, copyrights, and other proprietary measures. This remains true today for many software programs such as Microsoft Windows™, Apple iOS™, and others. But, Linux is an exception. It is an outstanding example of *open source software* (OSS).

OSS is a modern movement away from the traditional proprietary software development model to a radically different one. The goal is to enable all who are interested and willing to participate in the development, improvement, testing, and debugging of a piece of software as well as its integration with other software. The approach can reduce the investment burden on any single entity by tapping into the collective effort, power, and wisdom of many other developers in online communities. Ultimately, society benefits from the fruits of labor from these communities.

> **CT: Promote Free and Open**
>
> *Free and open can be an attractive alternative to proprietary and closed.*

Major initiatives are the *Open Source Initiative* (OSI) and the *Free Software Movement*. With some small differences, the two models are practically the same in that they make software free and their source code available for everyone interested to use, test, modify, improve, derive, deploy, and distribute without undue restrictions. The two are widely referred to together as *FOSS* or *FLOSS* (Free/Libre Open Source Software).

The *free and open* approach has proven itself and produced many outstanding software systems, including Linux and Android (OS), Apache (Web server), Mozilla (Firefox) and Google Chrome (Web browser), **gcc** the Gnu C and C++ compiler, **mySQL** a relational database system, and more. This is not to say that all software should be FLOSS. On the contrary, free and proprietary software each has a role to play in the software marketplace. Open source software does have a unique quality—the source code can be studied by experts in an attempt to ascertain its reliability and security.

In the digital age, we don't want to underestimate the power of free information flow and the energy of motivated online communities. Their influence is not limited to the programming industry. For evidence, we need not go far beyond Wikipedia.

4.4 Graphical User Interface

Before the days of pixel-based color graphical displays, computers used character-oriented monochrome cathode Ray cube (CRT) for display (Figure 4.1). In those early days (1960–80), users interacted with their computers through the command-line interface (CLI) (Section 4.7). Modern computer monitors are high-resolution, full-color, bitmapped displays. The color of each pixel (picture element) is directly controlled by a group of bits (usually 24 bits) in the display's video memory. The speed and responsiveness of modern bitmapped displays underlie *graphical user interfaces* (GUIs).

A GUI uses icons on a *virtual desktop* (Section 4.5) and a pointing device (mouse or touch screen, for example) in addition to the keyboard, to interact with the user. Xerox Palo Alto Research Center developed the Alto personal

FIGURE 4.1 A Monochrome CRT Monitor

computer in 1973. The Alto was the first computer to demonstrate the *desktop metaphor* and GUI. Later (1981), Xerox developed the Star workstation, introducing many of the features commonplace in today's personal computers. However, Star was not commercially successful. The first effective GUI on an affordable personal computer[2] was introduced by the Apple Lisa (1983).

4.5 Desktop Overview

After login on your computer, the first thing you see is the *desktop*, from which you can launch programs, manage application windows, control the configuration of your computer, and perform many other tasks. A desktop provides a GUI to make operating your computer more intuitive through a *desktop metaphor* by simulating physical objects. Overlapping windows can be moved and shuffled like pieces of paper. Buttons (icons) can be pushed (clicked) to initiate actions.

Most operating systems supply a fixed desktop. Linux distributions, on the other hand, offer a number of alternative desktops with a high degree of user customization. A good understanding of the desktop and how to use it effectively can make life on your computer much easier. desktops on modern computer systems work in very similar ways. We will describe a typical desktop next.

4.5.1 Desktop Components

A typical desktop displays the following components:

- *Root Window*—After login, the entire graphical display screen is covered by the *root window*, which displays the desktop. It is the space where all

[2]Cost about $10K in 1983.

other GUI objects (desktop components and application windows) are placed and manipulated. The root window is also the *parent window* for all other windows.

- *Taskbar*—Usually in the form of a horizontal bar along the top, bottom, or side edge of the root window. The taskbar displays icons, called *launchers* to start/resume specific applications with a single mouse click. You can add/remove launchers on the taskbar. On Mac OS X, it is known as the dock.

- *Start Button*—Often in the form of a logo at the end of the taskbar. Clicking the start button exposes the *start menu*, which is the main gateway to your computer's applications, folders, and settings, as well as restarting and shutdown.

- *Desktop Objects*—On the desktop, you can place icons representing objects, such as files, folders, and application launchers, for easy access. Clicking (usually double clicking) an icon opens the associated program or folder. Drag (depress the left mouse button without releasing) an object to move its location on the desktop or into the trash bin. Right click an icon to select possible operations on it.

- *Application Windows*—An application will usually displayed its own window for user interaction. Multiple application windows can overlap. Thanks to the window manager, you can change the *input focus* from one window to another, as well as move, resize, maximize, minimize, unmaximize, or close each window as you like.

- *Workspace Switcher*—An OS often enables you to work with multiple *workspaces* and to switch your screen display from one workspace to another. A *workspace* is essentially a duplicate root window to provide more space for placing/grouping additional application windows. With several workspaces, you can spread out your application windows for easier use.

- *Notification Area*—Part of the taskbar is used to display alerts and messages from various applications. Clicking on a notification display usually brings up the application window, and right clicking on the icon reveals a menu of operations.

In addition, the desktop also supplies a hidden *clipboard*, which is a temporary data buffer, accessible to most/all programs, to support *copy/cut and paste* operations that transfer data from one place to another or between two programs. A cut/copy operation stores new data onto the clipboard, erasing its previous content.

Also, pressing the PRINT SCREEN key saves the display on your screen to the clipboard, ready to be pasted into a file or an image processing application.

While the underlying GUI concepts apply generally, their realization on touch-screen devices can be somewhat different than on traditional mouse-keyboard computers. The intuitiveness of the desktop lets you perform most tasks with hardly a learning curve. However, if you can extend beyond your comfort zone and put in the effort to learn more about your arena, you'll become much more effective and efficient. In time, people may mistake you for an expert!

> **CT: KNOW YOUR ARENA**
>
> *It pays to get to know your arena well, be it your operating system, school, workplace, or home.*

Take your workplace, for example. Do you know all the emergency exits? Location of the first aid kit (and what's in it)? Where the fire alarm triggers and/or extinguishers are? Access and application of portable defibrillators? Satisfy your curiosity and expand your comfort zone; it may just save someone's life, even your own.

4.6 Are You Talking to Me?

On a computer, you'll soon find yourself happily working with several applications simultaneously, each in its own window. And that makes you feel good and productive. Looking a bit beneath the surface, all this will become a little more complicated than it seems at first. Recognizing such complications can give us a deeper understanding and help us better navigate our way on the computer.

The situation is rather like coming into a room (the computer and its desktop GUI) and finding a top manager (the OS) and several employees (applications you stared) each speaking a different language (each application's own way of interacting with users). When you say something (with keyboard or mouse for example), the first question is, "To whom am I talking?"

Well, that is a good question. And the answer lies in *input focus* and *event handling*.

4.6.1 Input Focus

With a GUI, the user controls multiple windows (the desktop and application windows), each of which can be affected by a user input action (an *event*). But not all windows are interested in all events. A *mouse event* (moving the mouse cursor or using a mouse button) is reported to the windows containing the mouse cursor, namely the cursor-containing window and its parent windows.

CT: Pay Attention to Details

Train your eagle eye. Notice small and subtle differences. Don't forget the big picture, but always mind the details. The devil is in the detail.

Keyboard events (pressing and releasing a key) are reported only to the window that has the *input focus.* Normally, only one window has input focus at any time. You switch input focus by moving the mouse cursor to a different window (and clicking, if necessary). Often, a window changes appearance when it gains/loses input focus to provide a visual cue. Have you noticed such visual cues? Be sure the intended window has input focus before you type on the keyboard. Otherwise, the keyboard events may go to a different window or get lost altogether.

Sometimes, an application would pop up a dialog window to collect user input or get user confirmation in order to proceed. The dialog window could seize input focus and not let go (being *modal*). This is a way to insist that the user takes care of the input before performing other tasks. If a modal dialog gets covered up by another window, the user may feel that the system has frozen. When this happens, minimizing windows to find the dialog is often the solution. See **Demo:** `FocusDemo` at the CT site.

4.6.2 Event Handling

A GUI environment allows the OS and application programs to provide an intuitive user interface for users. GUI applications are event driven. An event-driven program mostly waits for an event from the user to know what to do next.

Typically, a GUI application displays various *GUI components* for users to manipulate. Common components include buttons, icons, scroll/slide bars, input fields, checkboxes (pick several), radio buttons (pick one), and pull-down menus. Figure 4.2 shows some familiar components. Each component registers with GUI support the particular events it wants to monitor. For example, an OK button will monitor a mouse click event.

Event handling involves three players: the user who generates an event, the application that receives and reacts to the event, and the GUI system that delivers the event to the correct component.

When starting up, a GUI application sets up the monitoring and handling of certain events with the OS GUI support. When such an event occurs, the

FIGURE 4.2 GUI Components

application receives the event, reacts to the event, handles it quickly, and goes back to waiting for the next event.

Let's now turn our attention to how events get delivered to GUI applications. When an event is detected, event support typically performs these actions:

1. Determines which component is to receive the event, following both A and B. (A) The mouse location on the screen or the input focus. (B) Whether or not the component is monitoring the event. The receiving component is the *event target.*

2. Creates an event object, a piece of internal data, to represent the event.

3. Reports the event to the event target, usually by invoking an event-processing procedure in it.

Thus, by using the displayed GUI components and choosing from available menus, users can operate any GUI application without memorizing commands, making programs intuitive to use. However, in some occasions, a less elaborate operating environment can be more effective for experienced users, as we will see next.

4.7 Command-Line Interface

In addition to the GUI interface, most operating systems also offer a command-line interface (CLI) to interact with the user[3]. Both the GUI and the CLI eventually use the same set of OS-supplied system calls to perform tasks.

The program that provides the CLI is known as a *shell*, which normally runs inside a *terminal window* that simulates a character-oriented CRT monitor on your screen. A shell is much less demanding on computer resources and can be very efficient in certain situations, such as remote access to a computer and performing system admin and network management tasks on

[3]A mobile OS usually does not provide a CLI.

servers. Shells are not as user friendly and easy to use as graphical interfaces. On modern computers, both the GUI and CLI are available.

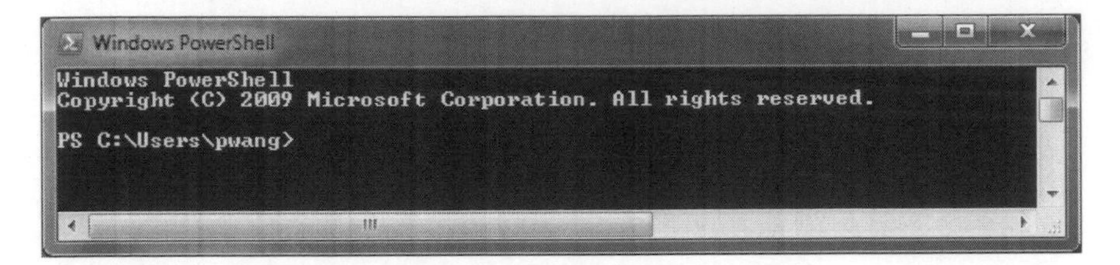

FIGURE 4.3 Windows PowerShell

Common shells include

- Microsoft Windows Command Shell and PowerShell (Figure 4.3)—Started by running `cmd.exe` or `powershell.exe`.
- Mac OS X Terminal Shell—Started by running the `Terminal` application.
- Linux BASH Shell—Started by running any terminal window.

A shell serves as your command interpreter and continually executes the *command interpretation cycle* (Figure 4.4):

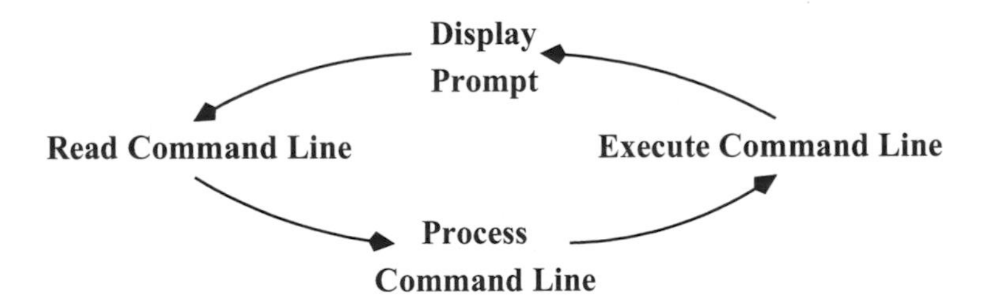

FIGURE 4.4 Shell Command Interpretation Cycle

1. Displays a prompt
2. Enables the user to type, edit, and enter the next command line
3. Breaks the command line into tokens (words and operators) and performs well-defined *shell expansions*, transforming the command line
4. Carries out (by calling shell-level functions) or initiates (by starting external programs) the requested operations
5. Waits for initiated operations to finish
6. Goes back to step 1

A command has the general form

command-name options ... files ...

The command-name identifies an application program or a function internal to the shell. The options control details on how it should operate. The files are usually the subject to be operated on. To experience the CLI, try **Demo:** `CLIdemo` at the CT site.

A shell provides a complete set of commands to perform any task you desire including:

- Launching applications
- Navigating folders and managing files
- Changing system settings
- Controlling applications that are running

For example, the command

```
rm -i 2009*.jpg
```

removes (deletes) any and all pictures, in the current folder, whose names start with `2009` and end in `.jpg`. The magic shell character `*` here stands for "any sequence of zero or more characters." The option `-i` says, "ask the user to interactively confirm the action before deleting a file." This particular command should work in BASH Shells (Mac OS X and Linux). The equivalent command

```
remove-item -Confirm 2009*.jpg
```

works under Windows PowerShell.

With a CLI, a simple one-line command can find files containing certain words or phrases on your computer. Imagine how hard that is using the GUI.

A shell allows you to start multiple programs running concurrently, switch your attention from one to another, suspend and resume each at will, and force task termination when you want.

In a shell, you can issue multiple commands on a single command line, connect the output of one command to the input of a second command (known as pipelining), and reuse previously issued commands recorded in a *history list*. Furthermore, you can write *shell scripts*, programs involving a series of commands, perhaps in loops and controlled by logical conditions. Shell scripts can become new commands and therefore automate frequent tasks and avoid having to repeat the same steps manually every time, as required in a GUI.

When accessing a computer system remotely (from a computer where you are, the local system), it makes sense to use a shell rather than the full GUI on the remote system. This is because graphical display and event transmission across a network is costly and slow. An unresponsive GUI is not fun, to say the least.

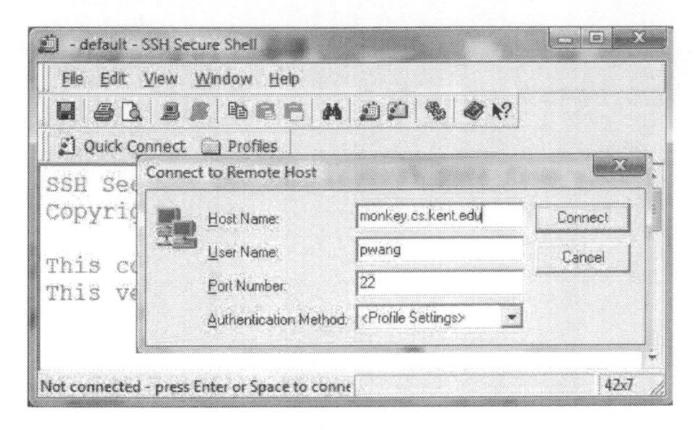

FIGURE 4.5 SSH

Common remote shell applications include SSH (Secure Shell; Figure 4.5) and Windows Remote Shell. They allow you to access a remote computer, log in, then start using the remote system through its command shell.

> **CT: MIND THE TRADE-OFF**
>
> *You gain some and you lose some. What's best often depends on the time, place, people, and the metric.*

The automobile gets you there fast in comfort but has negatives on resources and the environment. The bicycle is more effort, less comfort, but very energy efficient and great exercise. A feature-laden digital camera is wonderful if you use all those settings, but it can be expensive and bulky. A simple camera is much cheaper and easier to carry, but may not have the features you need upon occasion.

In computing, the GUI is engaging, user friendly, and intuitive. Without the GUI, personal computers may not have succeeded. Yet, the CLI can be much more efficient, powerful, and amenable to programming. It is especially good for accessing computers remotely. The GUI is demanding on computer resources, while the CLI is not.

In a real sense, life can be viewed as a sequence of choices. Make good choices and be aware of the trade-offs.

4.8 Files

A computer file is a sequence of bytes to encode and represent data for some particular type of information such as a Web page, a piece of email, a picture, a song, a movie, a presentation, a tax return, and so on. A critical OS function is to organize the storage and retrieval of files on your computer. Files are usually stored in *nonvolatile memory*, such as a HHD, SSD, or USB flash drive. Files are easily uploaded and downloaded via computer networks.

Well-designed encoding and format for files are used to represent different *information contents*. A file can only be handled by application programs that understand its encoding and format. For example, `.doc` files are formatted documents by MS Word, `.jpg` and `.png` are raster image files for pictures and graphics, `.mp3` files are for songs and music. File formats may be *open* with publicly available specifications or proprietary when the right to use such file formats are restricted by vendors.

4.8.1 File Content Types

The Internet standard MIME (Multipurpose Internet Mail Extensions) defines file *content types*. The content type information, not being part of the file content, may not be kept inside a file. Yet, it is critical that a receiving application knows the content type in order to properly handle the file.

TABLE 4.1 Content Types and File Suffixes

Content Type	File Suffix	Content Type	File Suffix
`text/plain`	`txt`	`text/html`	`html htm`
`text/css`	css	`application/javascript`	js
`application/pdf`	`pdf`	`application/msword`	`doc, docx`
`image/jpeg`	`jpeg jpg jpe`	`audio/basic`	`au snd`
`audio/mpeg`	`mpga mp2 mp3`	`application/x-gzip`	`gz tgz`
`application/zip`	`zip`	`audio/ogg`	`oga, ogg`
`video/ogg`	`ogv`	`video/webm`	`webm`

There are hundreds of content types in use today. Many types are associated with standard *file name suffixes* as shown in Table 4.1. A file suffix forms the trailing part of a file name. When viewing a listing of files in a folder, the file suffix may or may not be displayed. Often, instead of the suffix, a GUI may display a small icon indicating the default application that will open the particular file. Examining the file's properties should reveal the suffix and other attributes, such as file size and creation date.

The MIME content type will accompany any file attached to an email or retrieved on the Web.

4.8.2 File Tree

An operating system manages the systematic storage and retrieval of files. A file *folder* may contain individual files as well as folders. A folder is a special file, called a *directory*, that stores the names and attributes of files (including folders) contained in the folder. Files are stored in a hierarchical structure known as a *file tree*. The top-level folder of a file tree is the *root directory*. Other files and folders are stored, directly or indirectly, under the root folder.

On Mac OS X and Linux systems, the root directory is designated by the single character / (a SLASH). On Windows, a BACKSLASH is used instead.

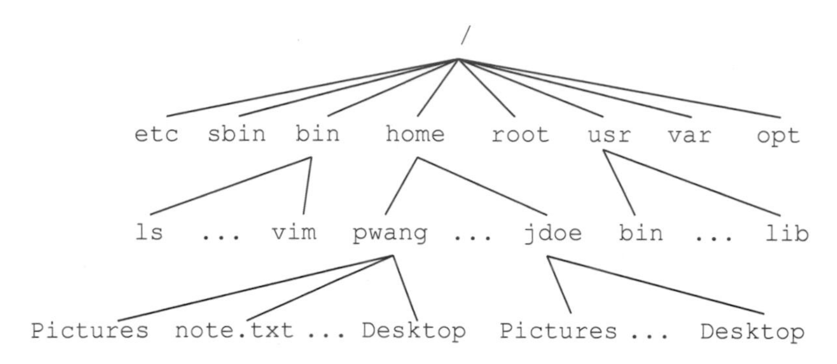

FIGURE 4.6 A File Tree

Figure 4.6 shows a typical Linux file tree. Other operating systems use basically the same structure for storing files. OS-related program files, configuration files, system settings, are placed in system folders. For example, on Linux or Mac OS X, `/bin` and `/usr/bin` contain application programs. On Windows, executable programs are placed under `C:\Program Files`.

Often, each user is given a *home folder* under which all files belonging to that user are kept. On Windows systems, the home folder is usually `C:\Users\`*userid*. On Linux systems, the home folder is usually `/home/`*userid*.

To access a file or folder on the file tree, you can either manually navigate to it on a GUI or specify its file location. There are several methods to specify the file location. The most general, and also the most cumbersome, is to list all the nodes in the path from the root, through a sequence of subfolders, to the file or folder you want. This path, which is specified as a character string, is known as the *absolute pathname*, or *full pathname*, of the file. On Linux and Mac OS X, after the initial /, all components in a pathname are separated by the character /. For example, the file `note.txt` in Figure 4.6 has

```
/home/pwang/note.txt
```

as its absolute pathname.

MS Windows uses *drive:*\ for root (for example, `C:\`) and \ (the BACKSLASH) as the separator (for example, `C:\Users\pwang\note.txt`).
The full pathname is the complete name of a file. As you can imagine, however, this name often can be lengthy. Fortunately, a filename also can be specified relative to the *current folder*. Thus, for the file `/home/pwang/note.txt`, if the current folder is `/home`, then the name `pwang/note.txt` suffices. A *relative pathname* gives the path on the file tree leading from a subfolder to the desired

file. A simple filename can be used when the current folder contains the file directly. Thus, a filename has three forms

- A full pathname (for example, `/home/pwang/note.txt`)
- A relative pathname (for example, `pwang/note.txt`)
- A simple name (for example, `note.txt`)

Your OS provides a file browser with which you can display, navigate, and manage files on the file tree. For example, the Windows File Explorer or Macintosh HD is such a tool.

Storing files in a tree structure provides a good hierarchical organization and easy access from a starting point (the root) to any files contained in the tree. There is also room for expansion to grow the tree. It is easy to remove parts of the tree by pruning. Furthermore, any branch of the tree has the same exact organization and structure as the whole tree. The recursive nature is important because an algorithm for any part of the tree should work for any other part, or the entire tree.

CT: Learn from Trees

A tree, with its recursive structure, is an excellent way to organize objects in a hierarchy that afford flexibility, efficiency, and scalability.

Other than the plants themselves, tree structures are actually commonplace: postal addresses, government offices, genealogy information, management of companies, military commands, and book contents. Such structures are all characterized by the fact that when you zoom in on a subpart, it resembles the whole thing, namely, they are recursive.

4.8.3 File Management and Access Control

Operations on files include

- Creating a new file or folder
- Deleting an existing file or folder
- Renaming a file or folder
- Replacing the contents of a file
- Copying files/folders

- Moving a file/folder to a different location on the file tree
- Setting access restrictions on files/folders

Because these operations must be performed by the OS kernel, system calls are provided for applications to request such file operations.

For security and privacy reasons, operating systems provide access control for files and folders. Each user is given a unique user ID and may belong to one or more groups of users who wish to cooperate on the computer. When a user runs an application, that program takes on the user ID and group IDs of the user. When a file is created, it is assigned an owner ID (owner of file) and group ID (group of file). The file owner can set the file group ID as appropriate. Usually only a super user or system administrator can change a file's owner. When a program attempts to access a particular file, the following pieces of information allow the OS to decide whether to grant or deny the access.

- The user/group ID of the executing program
- The user/group ID of the file being accessed
- The access permission settings of the file
- The intended access by the program (read, write, or delete for example)

For example, a regular user usually cannot modify or delete files belonging to another user. But, a system administrator can. Team members in a project may store files with a common group ID to allow easy sharing among group members. By setting the desired access permissions, users can control how/if certain operations on files are permitted and by whom.

4.9 Processes

On a computer, many individual programs are usually running at the same time. For example, you may have a Web browser, a music player, a document processor, and an email client running simultaneously. In addition to such applications that are visible to the user, there are also hidden applications running all the time, such as the window manager, the firewall, the time server, and the chat/message server, just to name a few.

In an operating system, the term *process* or *task* refers to a program (an application) that has started execution but has not finished.

A key OS kernel service is *process control*, coordinating and managing these *concurrent processes.*

Because a process must occupy the CPU to run, there can only be as many true simultaneous processes as the number of CPUs on the computer. The OS keeps concurrent processes running by rapidly switching the available CPUs among the processes. The technique is known as *preemptive multitasking*, which is used by all modern operating systems. What happens is that the

OS suspends a running process after a predetermined time period, called a *time slice*, and reassign the freed CPU to another process. Typically, a time slice can range from a few milliseconds (ms) to 100 ms or so.

When a process is suspended, the OS must save its current execution state, the *process context* of the process being suspended, so that the OS can later restore the state to resume the process. A process context is nothing but a collection of attributes and their values, allowing the OS to recreate the exact execution environment when a process was suspended.

> **CT: Keep It in Context**
>
> *Always keep information in its proper context. Make a point to preserve the context in order to properly interpret data.*

Contexts are basic to communication. Ever call customer service and had to repeat your name, phone number, address, and so on multiple times to service representatives? Was it not frustrating? But, we do understand because they need to place your call in the correct context. You just wish they would transfer your call together with its context so you don't have to repeat yourself so many times.

These days, we have multiple modes of communication: face to face, telephone, email, texting, online chat, and so on. There are pros and cons. A face to face is direct, immediate, and allows nonverbal expressions. A phone call can be direct if you reach your target. Later, when you follow up by phone, make sure you restore the conversion context first. Email is not direct but allows people to give considered replies. Plus the context is automatically kept by including a history of the exchanges in each reply. Texting and online chat are usually more mobile and immediate than email. Yet, they may have length limitations and provide no easy way for attachments.

In any case, selecting the proper communication mode is important, and paying attention to the context to avoid misunderstanding is always a must.

We may think of multitasking this way. A busy restaurant has many demanding patrons (dinning processes that insist on being served at once) but few tables (CPUs), and one waiter (the OS). The waiter can perform preemptive multitasking by moving patrons in and out of tables (context switching), and serving each table in quick succession. If it is done well, the waiter can keep all patrons happy. But the poor waiter must be so exhausted!

Thus, for multitasking, the OS:

1. Suspends a running process when its time slice ends or when it needs to wait for a certain condition, such as I/O completion, before proceeding

2. Saves its process context
3. Brings back the process context for the next process to be executed.
4. Resumes execution (of the next process)

The procedure is known as a *process context switch*

CT: Capture the State

Be aware of the state concept. Recording all critical parameters and their values of a system or situation enables you to save its state, making it possible to recreate the system or situation.

Recording the positions of all pieces and who made the last move saves the state of a chess match. Photographs and descriptions save the state of a crime scene. Detailed medical history saves the state of your health. Keeping track of your income, expenses, debt, taxes, and assets gives a good picture of your financial state.

Each individual has a physical state: height, weight, blood type, sex, eye color, hair color, and so on. It may also include blood pressure, heart rate, allergies, vaccinations, and surgical history. Keeping track of such information is important for people. *Electronic Health Records* (EHR) can help and are increasingly used in many parts of the world.

In movie making, every detail of a scene must be recorded in case action needs to resume at a later shooting or in a different location. Such details, (shall we say *scene state*?) help maintain film continuity.

Process context switching is just one concrete example of state saving in computing that enables multitasking in operating systems.

4.9.1 Process Lifecycle

To manage all processes, the OS uses a systemwide *process table*. Each process is represented by an entry in the process table that records important parameters for the process, including its *status*. A process usually goes through a number of *statuses* before running to completion. Figure 4.7 shows the process lifecycle.
The status of a process can be one of

- *Running*—The process is executing.
- *Blocked*—A process in this status is waiting for an *event* to occur. Such an event could be an I/O completion by a peripheral device, the

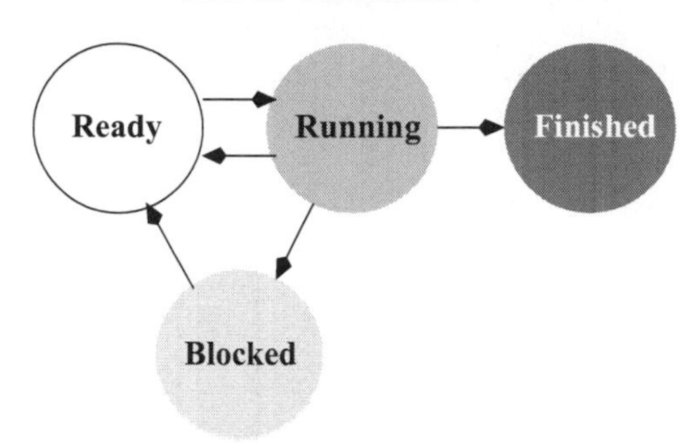

FIGURE 4.7 Process Lifecycle

termination of another process, the availability of data or space in a buffer, the freeing of a system resource, and so on. When a running process has to wait for such an event, it is *blocked* and waiting to be unblocked so it can continue to execute. A process blocking creates an opportunity for a *context switch*, shifting the CPU to another process. Later, when the event, for which a blocked process is waiting, occurs, the process awakens and becomes *ready* to run.

- *Ready*—A process in this status is then scheduled for CPU service. When a process is suspended at the end of its time slice, its status changes from running to ready.
- *Finished*—After termination of execution, a process goes into the finished status. The process no longer exists. The data structure left behind contains its exit code (success or failure) and any timing statistics collected. This is always the last status of a process.

A process may go through the intermediate statuses many times before it is finished.

4.9.2 Process Address Space

A program must be rendered in executable machine code (Section 10.5), instructions and data in binary, before it can run on a computer. For example, the high-level statement

```
if ( a > b ) then function_x(),
```

when translated to machine code, becomes something like the following:

1. Take value stored at memory address 1024 (to get `a`)
2. Take value stored at memory address 1028 (to get `b`)
3. If the value `a-b` is positive, then go to memory address 262144 (to run `function_x`)

Machine code instructions are executed one after the other. However, certain operations can make execution go elsewhere. As you can see from the example, instructions refer to memory addresses all the time. Thus, machine code is created relative to a preassigned contiguous memory space known as the *address space*.

4.9.3 Virtual Address Space Layout

The machine code for a process, residing in secondary storage (hard disk), must be brought to RAM before it can be executed by the CPU. To run many processes concurrently and have them share the available RAM, we cannot give processes real RAM addresses as their address space. Instead, we tell each process that its address space always starts from 0 and ranges up to a certain given limit. This is the *virtual address space* for a program.

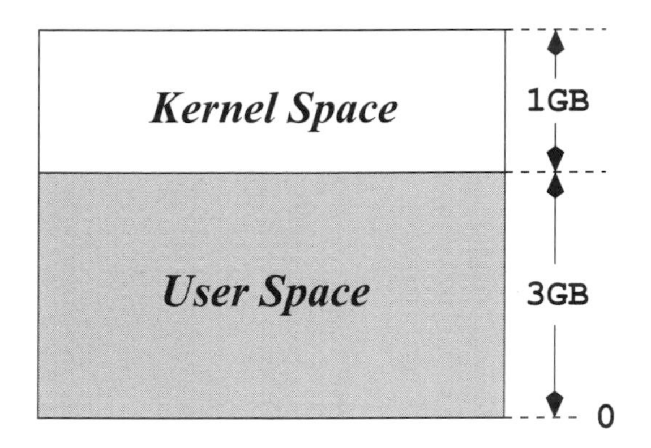

FIGURE 4.8 Virtual Memory Layout of a Process

Figure 4.8 shows a typical layout of the virtual address space.

- The *kernel space* is occupied by the OS kernel shared by all processes.
- The *user space* is occupied by an application's machine code.
- The sizes of the user and kernel space are constants for any particular OS. On a 32-bit system, the address space is almost always 4GB with a 1-2GB kernel space. On a 64-bit system, the address space is usually 256 TB.

A process executing instructions in user space is in *user mode* and has no access to the kernel space except through system calls provided by the kernel. Upon a system call, control is transferred to a kernel address and the process switches to *kernel mode*. While in kernel mode, the process has access to both user space and kernel space. The process switches back to user mode upon return of the system call.

4.9.4 Address Mapping

The OS prepares a process for execution by creating and storing its machine code in secondary memory following the virtual address space layout. In order to be executed by the CPU, data and instructions must be brought from virtual space into RAM. *Address mapping* is a technique to dynamically move chunks of machine code from virtual space to RAM for execution. Address mapping is usually implemented by the OS with hardware assist for speed. An address translation hardware in the CPU, the memory management unit (MMU), automatically translates virtual addresses to actual RAM addresses. Address mapping effectively shares RAM among all concurrently running processes. Furthermore, application program size won't be limited by the RAM size.

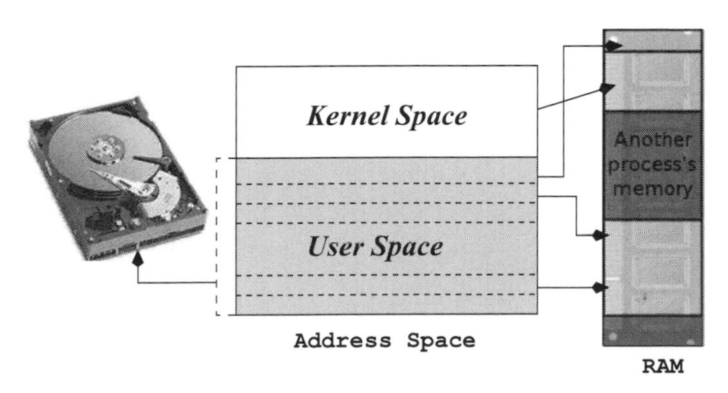

FIGURE 4.9 Virtual Memory Mapping

Here is an overview of the memory address mapping technique known as *paging* (Figure 4.9).

- The virtual address space is divided into chunks called *pages*. The RAM is divided into *page frames* to store the pages. Pages can be easily *swapped* into and out of RAM.
- A *page* is a block of contiguous virtual memory, usually at least 4 KB but can be larger. Pages may be of the same size or several different sizes.
- To start a process running, the OS brings an initial set of pages into RAM and executes the first instruction.
- Execution continues as long as virtual address references translate to actual addresses in RAM.
- When an instruction refers to a virtual address not in RAM, memory translation causes a *page fault*. The OS takes care of a page fault by

swapping into RAM page(s) needed to allow the process to continue execution.

- To swap in a page, a free page frame is used. When all frames are occupied, a well-designed algorithm picks a suitable page frame to be swapped out. Writing the swapped page back to where it came from is only necessary if the page content has been modified.
- The OS maintains a *page table* for each concurrent process to record paging-related parameters.
- An MMU, which is special hardware in the CPU, uses the page table for speedy virtual to actual address translation.

CT: TIMESHARING

Allowing multiple entities to share the same coveted/expensive resource at nonoverlapping times can make everyone happy and optimize resource utilization.

In real life, hotels, timesharing vacation properties, and car rentals, just to name a few, work this way. We can say these are forms of virtual ownership.

The many processes on your computer, through multitasking and paging, timeshare the available CPUs and RAM to run virtually in parallel. And you wouldn't have it any other way.

4.10 Managing Tasks

As users, we all enjoy easy access to files and running multiple tasks concurrently. We carry on doing what we do on the computer without giving it too much thought. Well, that is the way it should be. Get the computer out of the way of computing.

But, from time to time, the system freezes and becomes nonresponsive to your actions. Most likely, an application you are running got stuck. Or, in geek speak, a program or the system *hangs* or is *hung*.

In such a situation, many users will reach for the power switch on the computer. Powering the computer off and on again reboots the computer and surely will fix the system hanging problem. But you will lose any unsaved work, and you need to start everything over again.

Before doing anything drastic, you ought to give the *task manager* a try.

A task manager is an application that provides information and control over programs running on your computer. Open a task manager to see applications your are running, all processes and their status, and performance statistics of your system. You can also forcibly terminate an application or process that you suspect is causing the system freeze. Or you can request a soft restart (reboot).

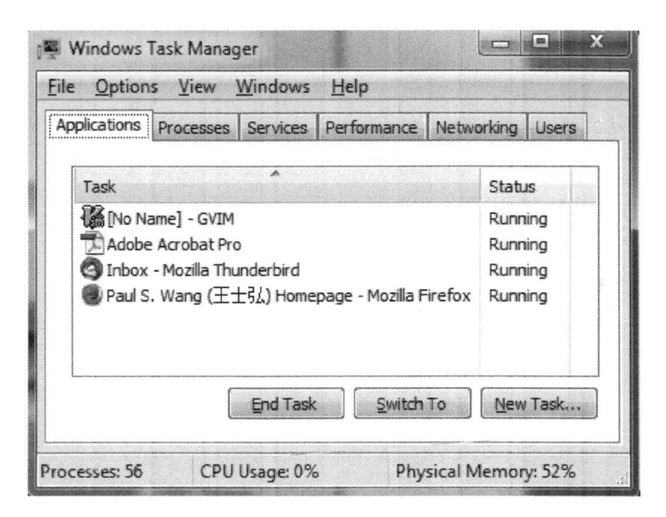

FIGURE 4.10 Windows Task Manager

Figure 4.10 shows the basic task manager on MS Windows. The Applications tab displays the user applications currently running. Presumably, one of them is hung, and you have a good idea which it is. Just select that application and click `End Task`, and you are unstuck.

But, how does one open the task manager if the system is frozen?

- To open Windows Task Manager—Use the key CTRL+SHIFT+ESC or CTRL+ALT+DELETE.
- To open Mac OS X Activity Monitor—Use COMMAND+SPACE to call up the spotlight box, enter and open "activity monitor," or use the Force Quit dialogue with COMMAND+ALT+ESC.
- To open a terminal window on Linux—Try CTRL+ALT+F2 for a terminal window and use the CLI to fix the problem. Use CTRL+ALT+F1 to get back to GUI. As a last resort, use CTRL+ALT+DELETE to reboot.

Play with the task manager, and you'll get more of a feel for what tasks and processes are all about. Then you'll be ready when a system freeze happens.

When a mobile device (smartphone or tablet) freezes up, usually your only recourse is to reboot. Simply hold the power button down for a short time and the device should reboot.

4.11 Up and Running

When up and running, the modern computer is such a powerful and flexible machine. But, even this machine has to go through initial stages of bringing itself alive. For a computer, performing the initial set of operations after power on is known as *booting*. The term *boot* is short for *bootstrap* derived from the common saying "pull yourself up by your bootstraps." The phrase conveys the general meaning of self-reliance and proceeding without external help.

Booting is a nontrivial procedure and will take a few moments to perform. Usually it involves these steps:

1. Hardware power-on self-test (POST)

2. Loading and starting an operating system

After power on, the CPU executes a program called the *bootloader*. The bootloader is stored in nonvolatile, read-only memory (ROM) or flash memory, making it a *firmware*. Flash-based firmware is *field-upgradable*, allowing user download and update of the firmware program.

A bootloader first performs POST to test and initialize hardware devices, including CPU, RAM, video display card, keyboard, mouse, and disk drives. POST is time consuming and only performed after an actual power on. It is skipped for a reboot.

After POST, the bootloader proceeds to find and load the OS kernel. Then execution transfers to the OS kernel, which continues to set up the full process, network, and user environments for the computer. A bootloader may allow the user to choose the device from which to load the OS or to choose from different operating systems. Thus, you can recover from a damaged system by booting from a backup system disk. Or you can choose between distinct operating systems. People with dual-boot Windows/Linux systems will tell you the advantage of such bootloaders.

The BIOS is a widely used bootloader on PCs. Starting in 2010, systems began to move from BIOS to the newer Unified Extensible Firmware Interface (UEFI), which is expected to replace BIOS.

Operating systems are still evolving, especially those for mobile devices. They will become more efficient, personalized, and self-adjusting to each individual.

CT: Better Control Better System

Improve, nourish, and perfect system control can make the entire system stronger, healthier, more effective, efficient, secure, and less prone to failure.

Just like a person with a different brain is a different person; a computer with a different OS is a different computer. We improve the OS to make better computers and, likewise, we should train our mind, our nervous system, to improve the person. Medical science has begun to promote rhythmic breathing, meditation, and yoga for improving our health and well being. Perhaps we are only starting to discover our own control system and its relation to the whole body. Hopefully, computers and networks can also be used in many new ways in businesses, organizations, and governments to improve their control systems for the benefit of all their members.

In summary, the OS gets to manage your computer soon after booting, provides a GUI and a CLI interface for users, organizes files, manages multiple tasks concurrently, and makes effective use of system resources. We should all appreciate what this master program does.

Exercises

4.1. What is a system call? A system program? What is the difference?

4.2. What is kernel mode? User mode? Mode switching?

4.3. What OS does your computer run? List five system programs supplied by your OS.

4.4. What is CLI? GUI? How do they compare?

4.5. What is input focus among windows? Within a given window?

4.6. What is an event? Event handling?

4.7. What is a shell? Name two different shells.

4.8. What is a process? What is multitasking?

4.9. What is virtual address space? Paging?

4.10. What is virtual address space? Paging?

4.11. Explain why shutting down or rebooting your system tends to fix things.

4.12. **Computize**: Describe the structure of a tree. Give real-life examples of tree structures.

4.13. **Computize**: What is context switching? Give real-life examples of context switching.

4.14. **Computize**: What are the trade-offs for young people between "Study and learn hard" and "Play games and hangout with friends"?

4.15. **Group discussion topic**: *State of mind.*

4.16. **Group discussion topic**: *Open source projects and the RepRap 3D printer.*

4.17. **Group discussion topic**: *This is how I multitask.*

4.18. **Group discussion topic**: *This thing about the computer system puzzles me.*

Chapter 5

Hello There!

"The net is down again, I am calling my ISP right now." "Oh, you don't have Wi-Fi here?" We all had such dreaded moments. The computer is a powerful machine. But without being connected and able to communicate with others, it is such an isolated and lonely existence.

Computers are heavily dependent on networking to do many basic things, such as software update, email, instant messaging, video call, surfing the Web, and even setting the system clock. Actually, it is hard to tell where the computer ends and the network begins. No wonder why people say, "The network is the computer."

Basically, processes on one computing device can communicate and interact with processes on other computing devices connected by networks. The *interprocess communication* is done by a combination of networking hardware, software, protocols, and OS support on heterogeneous computer systems.

The Internet spans the globe and brings all parts of the world within instant reach. It has indeed created for all humanity the "global village." An understanding of networking is important for all people, not just those who study computers.

5.1 What Is a Network?

A *computer network* is a communications medium that connects *hosts* and enables them to exchange data and share resources with great speed. A host may be a computer, a smartphone, a network printer, a network disk, a set-top box, or some other computing device. A network can be represented as a

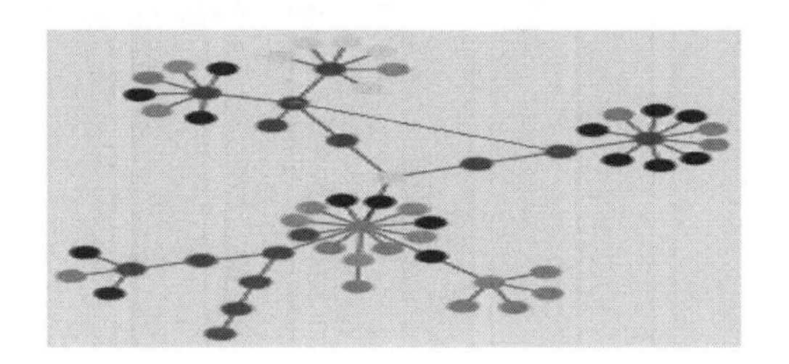

FIGURE 5.1 A Network Graph

graph, a structure consisting of nodes connected by edges (Figure 5.1).

- Internal nodes—Communication and routing processors
- Edges—Data transmission links
- Terminal nodes—End-user hosts

Internal nodes perform network functions and may include hubs, switches, bridges, routers, gateways, and so on. A large network has a great number of nodes. The Internet, the largest global network, has multiple billions of nodes at any given time. Small networks may use a simpler topology (connection geometry), where there may be no distinction between internal and terminal nodes, as shown in Figure 5.2.

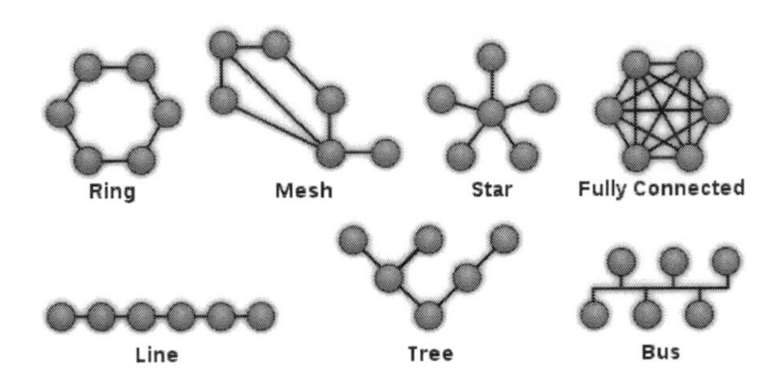

FIGURE 5.2 Network Topology

Services made possible by a network may include:

- Electronic mail
- World Wide Web, E-business, E-commerce, and social networks
- Online chatting and Internet phone calls
- Audio and video streaming
- File transfer
- Remote login
- Cloud computing

5.2 The Internet

Ever since the invention of the telephone, the world has been building electronic communication infrastructures. The desire to use the telephone's analog

FIGURE 5.3 Early Telephone Modem

network to transmit digital data was natural and obvious. The *acoustic modem* (modulator-demodulator; Figure 5.3) was invented for this very purpose. A user would connect a modem to a terminal and dial the number of a remote computer, then "talk" to the office computer from home!

How clunky! Today, digital networks—telephone (landline and mobile), TV, and streaming media—are merging into one big global infrastructure.

The Internet is a global network that connects computer networks using the *Internet Protocol* (IP). The linking of computer networks is called *internetworking*, hence the name Internet. The Internet connects numerous networks in homes, schools, businesses, and governments. Users of the Internet world-wide number in the billions.

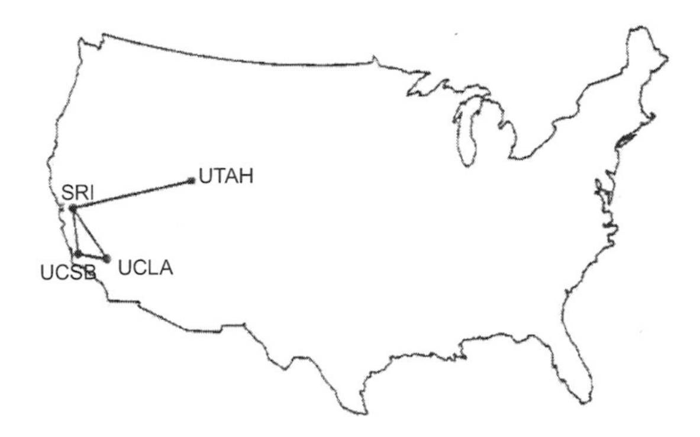

FIGURE 5.4 ARPANET MAP 1969

The Internet evolved from the ARPANET, a US Department of Defense Advanced Research Projects Agency (DARPA) sponsored research project for reliable military networking in the late 1960s. The experimental network started with four nodes (Figure 5.4): University of California Santa Barbara,

Stanford Research Institute, UCLA's Network Measurement Center and University of Utah.

ARPANET developed the IP as well as the higher level *Transmission Control Protocol* (TCP) and *User Datagram Protocol* (UDP) networking protocols. The architecture and protocol were designed to support a reliable and flexible network that could endure wartime attacks.

The transition of ARPANET to the Internet took place in the late 1980s as NSFnet, the US National Science Foundation's network of universities and supercomputing centers, helped create an explosive number of IP-based local and regional networks and connections. The Internet is so dominant now that it has virtually eliminated all historical rivals, such as BITNET and DECnet.

The *Internet Corporation for Assigned Names and Numbers* (ICANN, `www.icann.org`) is a nonprofit organization responsible for IP address space allocation, protocol parameter assignment, domain name system management, and maintaining root DNS servers (Section 5.11)

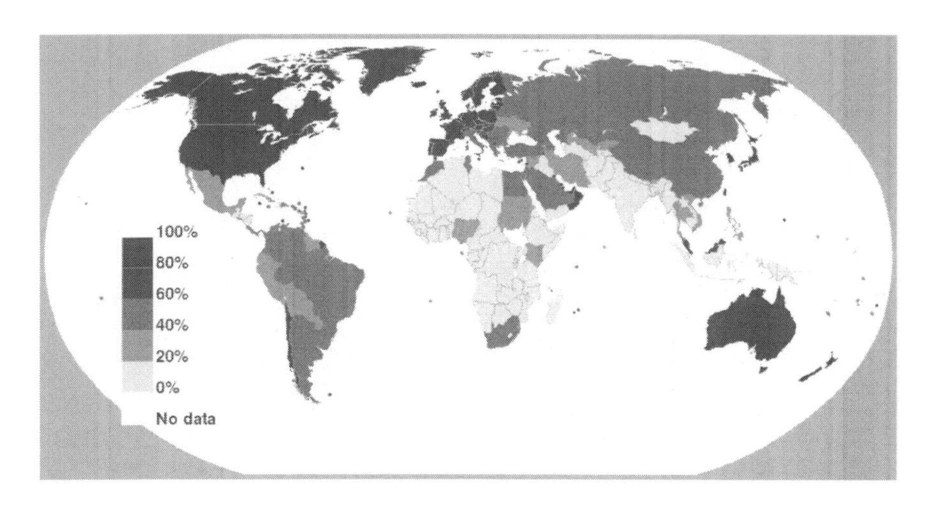

FIGURE 5.5 Modern Internet Usage Map

Figure 5.5 shows Internet users in 2012 as a percentage of a country's population (source: International Telecommunications Union).

5.3 Local and Wide Area Networks

The basic building blocks of the Internet are *Local Area Networks* (LANs) in houses, buildings, campuses, and other establishments. A LAN connects near-by hosts, normally well within 10 km. Most, if not all, LANs use the Ethernet protocol and run at 100 mbs (mega bits per second), 1 gbs (gigabit per second, 10^3 mbs), 10 gbs, and even 100 gbs. The speed, or *bandwidth*, of a network depends on the physical extent of the network as well as the data transmission hardware and software used.

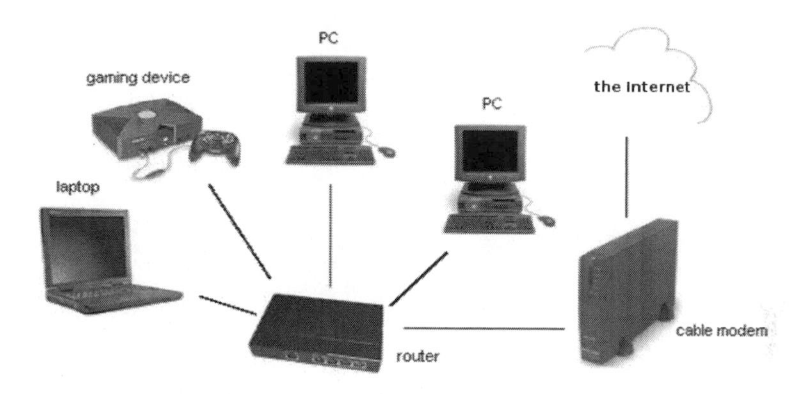

FIGURE 5.6 A Home Ethernet

Figure 5.6 shows a typical in-home Ethernet. Each host (PC, game console, printer, network disk drives) has a network interface card (NIC) that connects the host to a *router* through a standard (RJ45) Ethernet cable. Each Ethernet NIC has a unique vendor-assigned *media access control* (MAC) address, and the router relays network traffic across the LAN. A router usually performs these functions:

- Allowing hosts, as they are turned on/off, to be connected/disconnected dynamically to/from the LAN
- Assigning a distinct local IP address to each connected host
- Interfacing the LAN to the Internet
- Providing a firewall against unwanted Internet traffic

The router connects to the Internet, which is a *Wide Area Network* (WAN), via a modem supplied by an *Internet Service Provider* (ISP). The ISP can be a telephone company, a cable TV company, a satellite TV company, or some other telecommunications provider. Often, a router can also be a wireless access point allowing hosts to connect to the LAN using Wi-Fi (Section 5.5).

An Ethernet works as a *data bus*. A host sends a message by placing it on the bus. The message, called an *Ethernet frame*, contains the MAC address of its destination. All nodes on the bus can receive the message, but only the intended recipient accepts it. Larger Ethernet LANs may involve *hubs* to connect more hosts, *bridges* to extend the range of the LAN, and *switches* to enable more efficient direct communication between nodes. Large businesses operate *intranets* that connect LANs within the company using Internet protocols.

A WAN spans large distances and usually connects networks in different cities, regions, or countries. The Internet is, of course, the best-known WAN.

Typical WAN connection speeds people get at home, supplied by an ISP, range from 128 kbps (basic ISDN, *Integrated Service Digital Network*) to 150 mbps (high-speed broadband). Usually, upload speed (sending to the Internet) is much slower than download speed (receiving from the Internet). Upload speed can be just 10% of download speed.

5.4 Internet Architecture

The Internet consists of two parts:

- End-user LAN and hosts—Computing devices that are consumers of Internet services
- ISP networking devices—Specialized networking equipment and high-speed data links that collectively create and support the Internet infrastructure

The reliability and resilience of the Internet infrastructure result from its high degree of redundancy (multiple routes between nodes) and the fact that it neither needs nor allows central control or coordination.

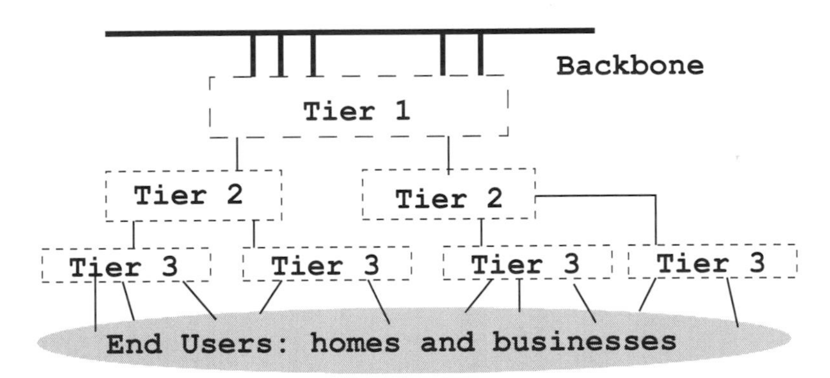

FIGURE 5.7 Internet Architecture

> **CT: Redundancy for Safety**
>
> *Provide redundancy and avoid loss of critical resources to gain system robustness and safety.*

The American saying goes, "Don't put all your eggs in one basket." And the Chinese saying goes, "The crafty rabbit has three dens (狡兔三窟)." For example, businesses, such as banks, will want to keep duplicate copies of their account records at multiple secure locations. It may require a bit more resources and complexity, but strategic redundancy can avoid system breakdown in the face of accidents or even deliberate attacks. At home, we all should back up important files to guard against possible loss.

Figure 5.7 shows the tiered Internet architecture where, generally, smaller tier 2 and 3 ISPs connect to larger tier 1 networks for delivery of traffic to destinations served by other ISPs. Tier 1 networks form the main Internet *backbone*. A typical backbone network uses fiber optic trunk lines, many fiber optic cables bundled together, for increased speed and capacity. Bandwidth between core nodes on a backbone can reach 100 gbs or more.

Tier 1 ISPs are usually the same companies that operate long-distance telephone networks. US tier 1 ISPs include AT&T, CenturyLink, Level 3 Communications, Sprint, Verizon, and Vodafone. Others in the world include Bharti (India), British Telecom, China Telecom, Deutsche Telekom AG, France Telecom, and Telefonica (Spain).

CT: ONE AND ALL

All for one, and one for all. It is the Internet!

With instant communication, information retrieval, and easy and affordable access, the Internet brought increased equality among people and continues to benefit humankind in profound ways.

The Internet is changing the way we communicate, work, play, obtain, and share information. It breaks space, time, and economic barriers and brings all more freedom to express, associate, create, and innovate. With the Internet, essentially all knowledge, expertise, news, opinions, and entertainment are just a few clicks away. Individuals everywhere can contribute their effort, artistry, and expertise with ease for all to share. And everyone can stand on the shoulders of the collective Internet giant and make further progress.

We are already seeing big impacts brought by email, e-commerce, e-government, distance learning, media streaming, online shopping, dating, social networking, Wiki, open source software, and so on. Still, the full potential of the Internet is yet to be realized.

5.5 Wireless Networking

A wireless network is generally not as fast or wide-ranging as a wired one, but can provide much appreciated mobility and convenience.

What is Wi-Fi?

Wi-Fi is a standard wireless LAN (WLAN). The IEEE[1] developed a series of 802.11 wireless communication protocols: 802.11b (up to 11 Mbps, 115–460 ft), 802.11g (up to 54 Mbps, 125–460 ft), 802.11n (up to 150 Mbps, 230–820 ft), and 802.11ac increasing speed to 0.87 Gbps. The n and ac standards also allow multiple data streams (MIMO) making them that much faster. These are collectively known as *Wi-Fi* (or WiFi), and generally providing faster speed and longer range in progression.

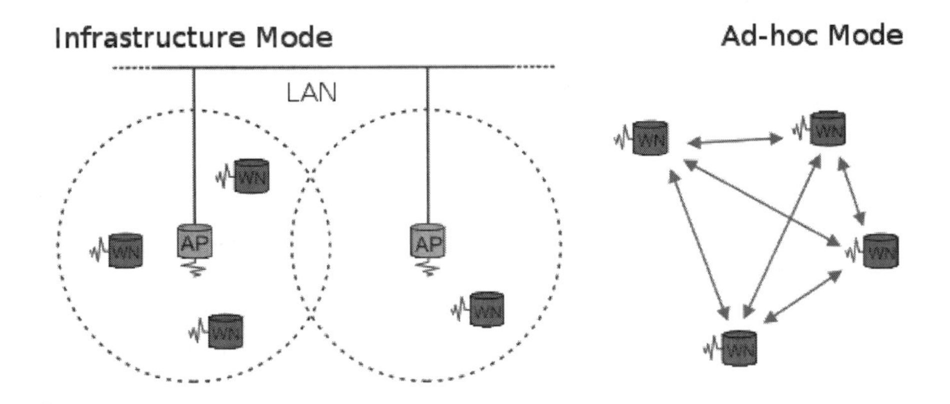

FIGURE 5.8 WiFi Network Modes

Wireless equipment, such as laptops, tablets, smartphones, set-top boxes, smart TVs, and others, usually communicate in *infrastructure mode* as wireless nodes (WNs) on a Wi-Fi network. Some are also able to communicate directly with a peer in *ad-hoc mode.*

- *Infrastructure mode*—Wireless nodes communicate with a central node, called an *access point* (AP), that in turn connects the WLAN to a wired LAN and the Internet.
- *Ad-hoc mode*—Two wireless devices communicate with each other directly peer-to-peer, without using an AP.

Most WLANs use infrastructure mode to access the Internet, a local printer, and other wired nodes. The ad-hoc mode supports wireless access from one device to another (Figure 5.8).

[1]Institute of Electrical and Electronics Engineers

Each wireless node requires a *wireless network adapter*, in the form of a wireless NIC or a USB adapter, to communicate wirelessly. A wireless router usually serves as an access point.

Bluetooth

Another well-known wireless standard is Bluetooth. Bluetooth technology uses low-cost radio chips that transmit at relatively low power and have a range of only 30 feet or so. Bluetooth networks also use the unregulated 2.4 GHz frequency range and are limited to a maximum of eight connected devices. The maximum transmission speed only goes to 1 mbps.

WiMAX

WiMAX (Worldwide Interoperability for Microwave Access), based on IEEE 802.16 supplies speeds of up to 10 Mbps. Operating around 2.5 GHz in the USA, WiMAX provides long-range wireless broadband from 10 to 30 miles.

Mobile Phone Data Service

Mobile phone networks provides digital communication and Internet access, using increasingly faster forms of data transmission: *General packet radio service* (GPRS), 3G, and 4G LTE. Cellphone networks compete to provide higher speeds and wider coverage for their data networks, which serve mainly smartphones and small tablets.

A computing device can be *tethered* to a smartphone or tablet with bluetooth, a USB cable, or Wi-Fi, so the tethered device can share the phone's data service to reach the Internet. In recent years, mobile networks have experienced explosive growth in many parts of the world.

5.6 Networking Protocols

For computers from different vendors, under different operating systems, to communicate on a network, a detailed set of rules and conventions must be established for all parties to follow. Such rules are known as *networking protocols*. Networking makes many different services available. Each networking service follow its own specially designed protocols. Protocols govern such details as:

- Address format of hosts and processes
- Data format
- Manner of data transmission

- Sequencing and addressing of messages
- Initiating and terminating connections
- Establishing services
- Accessing services
- Data integrity, privacy, and security

Thus, for a process on one host to communicate with another process on a different host, both processes must follow the same protocol. The *Open System Interconnect* (OSI) *Reference Model* (Figure 5.9) provides a standard layered view of networking protocols and their interdependence. The corresponding layers on different hosts, and inside the network infrastructure, perform complementary tasks to enable data exchange between the communicating processes (P1 and P2 in Figure 5.9).

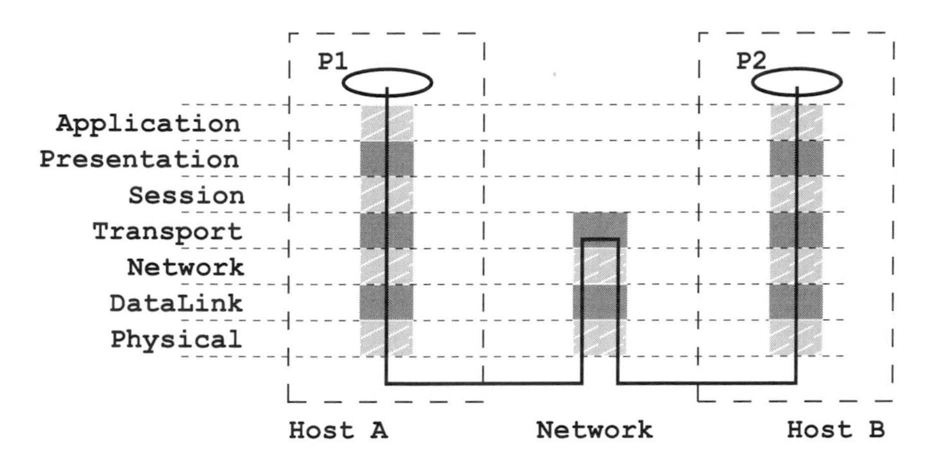

FIGURE 5.9 Networking Layers

Among common networking protocols, the Internet Protocol Suite is the most widely used. The basic IP is a *network layer* protocol. The TCP and UDP are at the *transport layer*. The HTTP (*Hypertext Transfer Protocol*) is at the *application layer* and is used for the Web.

CT: Follow Protocol

Only by following protocol can parties that may have never met cooperate smoothly.

Protocols can affect the effectiveness and speed of the entire system.

Networking protocols are no mystery. Think about the protocol for making a telephone call. You (a client process) must pick up the phone, listen for the dial tone, dial a valid telephone number, and wait for the other side (the server process) to pick up the phone. Then you must say "hello," identify yourself, and so on. This is a protocol from which you cannot deviate if you want the call to be made successfully through the telephone network, and it is clear why such a protocol is needed. The same is true of a computer program attempting to talk to another computer program through a computer network. The design of efficient and effective networking protocols for different network services is an important area in computer science.

5.7 IP Addresses

Every host on the Internet has its own network address that identifies the host for communication purposes. The addressing technique is an important part of a network and its protocols. An Internet IPv4 address is represented by 4 bytes in a 32-bit quantity. For example, `tiger`, a host at Kent State, has the IP address `131.123.38.172` (Figure 5.10). This *dot notation* (or *quad*

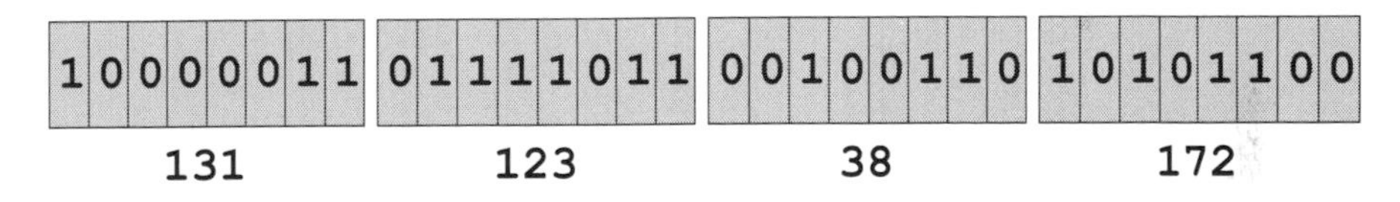

FIGURE 5.10 IPv4 Address

notation) gives the decimal value (0 to 255) of each byte. To accommodate its explosive growth, the Internet is moving to IPv6, which supports 128-bit addresses. The IP address is similar to a telephone number in another way: The leading digits are like area codes, and the trailing digits are like local numbers.

Basically, the Internet transmits information by routing *data packets* (Section 5.14) from a source IP address to a destination IP address.

5.8 Domain Names

Because of its numerical nature, an IP address is easy on machines but hard on users. Therefore, any host may also have an alternative address known as a *domain name* which is composed of words, rather like a postal address. For example, the domain name `tiger.cs.kent.edu` identifies a host `tiger` at the Department of Computer Science, Kent State University. Domain names are for convenience. It is not the case that every host must have a domain name.

With domain names, the entire Internet name space for hosts is recursively divided into disjoint domains in a tree structure (Figure 5.11), similar to a file tree (see Figure 4.6). The address for `tiger` puts it in the `cs` local domain,

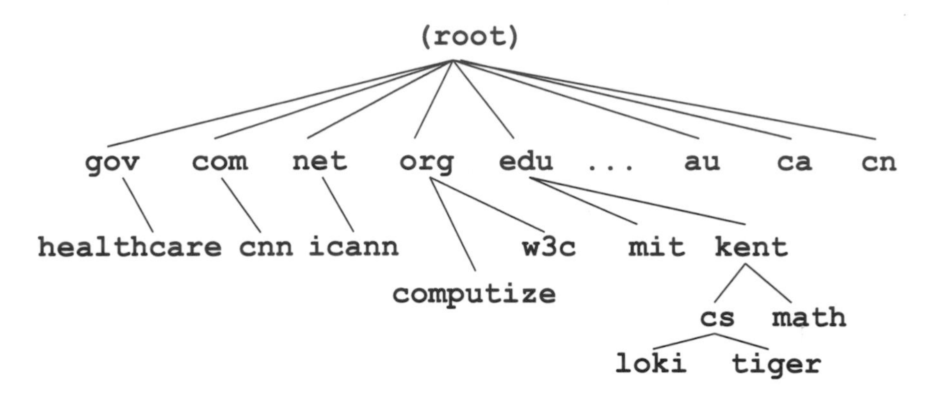

FIGURE 5.11 The Domain Name Tree

within the `kent` subdomain, which is under the `edu` *top-level domain* (TLD) for US educational institutions. Other TLDs include `org` (nonprofit organizations), `gov` (US government offices), `mil` (US military installations), `com` (commercial outfits), `net` (network service providers), `uk` (United Kingdom), `cn` (China), and so forth. By the way, our CT website has the domain name `computize.org`.

Hosts within a local domain (for example, `cs.kent.edu`) can refer to one another by their hostname alone (for example, `monkey`, `dragon`, `tiger`), but the full domain name must be used for hosts outside the local domain. Further information on Internet domain names can be found in Section 5.11.

All network applications accept a host address given either as a domain name or as an IP address. In fact, a domain name is first translated to a numerical IP address before being used to locate a host. For example, the Web host at Kent State has the domain name `www.kent.edu`, and the IP address `131.123.246.53`. You can visit the Kent State website using either the domain name or the IP address. Why not try it yourself?

Domain names are not created equal. Some are much better than others. People pay good money for great domain names that can help their businesses or causes.

5.9 Client and Server

A stand-alone application, such as Photoshop or WordPad, does not require a network connection. A network application, such as Firefox (Web browsing), Outlook (emailing), or Skype (video calling), won't work without one.

Most commonly, a network service involves a client and a server. The

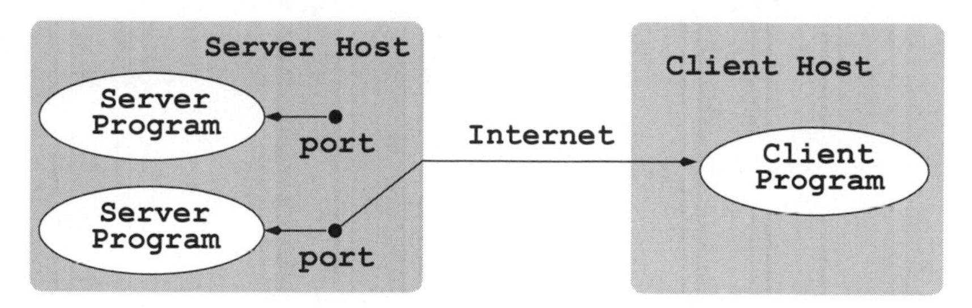

FIGURE 5.12 Client and Server

client–server pair of programs together provides the desired network service. The client application presents a user interface and acts as an agent for users to access a particular network service. The client alone is not able to fulfill a user's service request. It must initiate requests to a server program running on a specific host, and the server program responds by providing the needed service (Figure 5.12).

The computer on which the client application runs is called the *client host*, and the computer on which the server program runs is called the *server host*.

For widely used services on the Internet, clients and servers must follow standard protocols designed specifically for a particular service.

> **CT: INTEROPERATE**
>
> *By following the same protocol, a client can work with any server, and a server can work with any client.*

Clients for the same network service compete to offer features and functionalities for the end user. The multitude of available Web browsers demonstrates this clearly.

Similarly, different *server* processes are used to provide different kinds of network services. Servers from different sources may also provide the same service. The most obvious example is the Web. There are a great number of Web servers putting all the millions of websites online.

Common Internet client applications are:

- Email clients—Microsoft Outlook Express, Mac OS X Mail, and Thunderbird, for example, use Simple Mail Transfer Protocol (SMTP) to send and Post Office Protocol (POP) or Internet Message Access Protocol (IMAP) to receive mail messages from email servers.
- Secure remote host access clients—PuTTY and OpenSSH, for example,

use Secure Shell (SSH) protocol for remote login, and secure file transfer (SFTP) protocol for file transfer to communicate with SSH/SFTP servers on remote hosts.

- World Wide Web clients—Internet Explorer, Safari, Google Chrome, Firefox, for example, use HTTP and Secure HTTP (HTTPS) to access Web servers running on remote hosts.

Each *Internet standard service* has its own unique *port number* that is identical on all hosts. The port number, together with the Internet address of a host, identifies a particular server program (Figure 5.12) anywhere on the network. For example, `SFTP` has port number 115, `SSH` has 22, and `HTTP` has 80.

Nonstandard services can be proprietary or even private and can be assigned port numbers not in conflict with other well-known services. Anyone who is able may define a protocol for some purpose, implement client and server programs, and run them on various hosts on the Internet to communicate via the protocol. Or, standard services can be placed at arbitrary ports so the services are only available to those who know the IP and port numbers.

5.10 Peer to Peer

In client-and-server communication, the client and server are very different

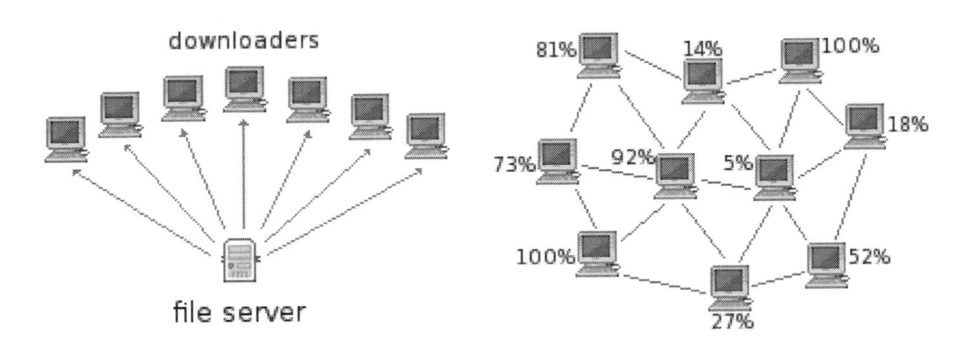

FIGURE 5.13 P2P File Sharing

programs with different capabilities. client and server are not equal. Only a client can initiate a service request. Only a server can respond to a service request. However, in *peer-to-peer* (P2P) communication, participating parties are equal have the same capabilities. They collaborate on equal terms and perform the same services for one another. Peers can request and provide services to one another.

The BitTorrent protocol for file sharing is a prime P2P example. Figure 5.13 left shows file downloading from a central file server, which can be slow and very taxing on the server. Furthermore, downloading fails if the server goes down. With P2P (Figure 5.13 right), files are cut up into many slices. Each peer uploads and downloads slices of a large file to/from other participating peers, sharing the work and the result alike. A bittorrent client joins a peer group to download a large file. After downloading a number of file slices from peers, it starts to upload those slices to other peers needing those slices. A peer having all file slices and ready to serve others is called a *seed.* Percentages in the figure show the various stages of file download completion for the peers.

5.11 DNS Service

As stated in Section 5.2, every host on the Internet has a unique IP address. A host often also has a domain name. The *domain name space*, the set of all domain names with their associated IP addresses, changes dynamically with time due to the addition and deletion of hosts, regrouping of local work groups, reconfiguration of subparts of the network, maintenance of systems and networks, and so on. Thus, new domain names, new IP addresses, and new domain-to-IP associations can be introduced in the name space at any time without central control.

The *domain name system* (DNS) is a network service to support the dynamic update and retrieval of information contained in the domain name space (Figure 5.14). Each local DNS domain (zone) runs its own DNS server(s) to supply information from locally maintained DNS databases. All DNS servers cooperate to respond to DNS data requests. Thus, the DNS is a distributed information system.

A network client (for example, a Web browser) will normally use the DNS

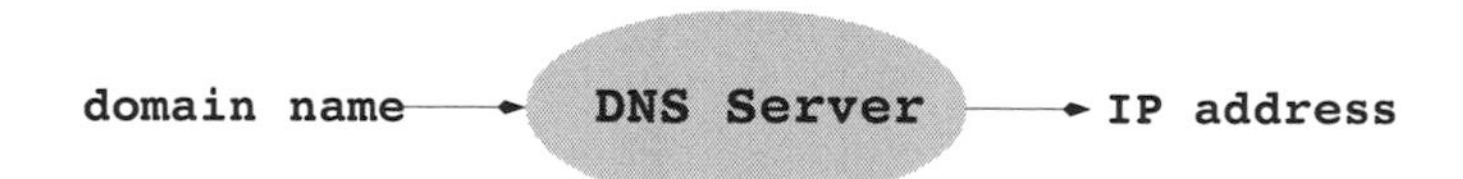

FIGURE 5.14 Domain to IP

to obtain IP address information for a target host before making contact with it. The dynamic DNS also supplies a general mechanism for retrieving many kinds of information about hosts and individual users.

Here are points to note about the DNS:

- The DNS organizes the entire Internet name space into a big tree structure. Each node of the tree represents a *domain* and has a label and a list of resources.
- Labels are character strings (currently not case sensitive), and sibling

labels must be distinct. The root is labeled by the empty string. Immediately below the root are the TLDs: `edu`, `com`, `gov`, `net`, `org`, `info`, and so on. TLDs also include country names, such as `at` (Austria), `ca` (Canada), and `cn` (China). Under `edu`, for example, there are subdomains `berkeley`, `kent`, `mit`, `uiuc`, and so on (Figure 5.11).

- A full domain name of a node is a dot-separated list of labels leading from the node to the root (for example, `cs.kent.edu.`).
- A relative domain name is a prefix of a full domain name, indicating a node relative to a domain of origin. Thus, `tiger.cs` is actually a name relative to `kent.edu`.
- A label is the formal or *canonical* name of a domain. Alternative names, called *aliases*, are also allowed. For example, if `tiger` also runs the local name server, then it may be given the alias `ns`. Thus, `tiger.cs.kent.edu` is also known as `ns.cs.kent.edu`. If the same host also runs the Web server, then it can be given another alias `www`.

CT: INDIRECTION ADDS FLEXIBILITY

Use an agent or representative to provide indirection and flexibility. It may be less direct, but you will gain a much needed degree of freedom.

Examples of indirection abound, for example:

- Retailers, wholesalers, and manufacturers
- Agents and movie stars
- Brokers and real estate owners
- Front-desk receptionists and company employees

In programming we use a variable `legal_drinking_age` instead of a fixed constant 21 directly. The indirection allows a procedure to work with any legal drinking age which may depend on laws in different localities.

For the Internet, domain names lead to IP addresses through the DNS system (the agent). The indirection allows meaningful, easy-to-remember domain names for users while enabling managers of server hosts to assign/change their IP addresses at any time.

5.12 DNS Servers and Resolvers

DNS Servers

Information in the distributed DNS is divided into *zones*, and each zone is supported by one or more *name servers* running on different hosts. A zone is associated with a node on the domain tree and covers all or part of the subtree at that node. A name server that has complete information for a particular zone is said to be an *authority* for that zone. Authoritative information is automatically distributed to other name servers that provide redundant service for the same zone. A server relies on lower-level servers for other information within its subdomain and on external servers for other zones in the domain tree. A server associated with the root node of the domain tree is a *root name server* and can lead to information anywhere in the DNS.

An authoritative server uses local files to store information, to locate key servers within and without its domain, and to cache query results from other servers.

The management of each zone is also free to designate the hosts that run the name servers and to make changes in its authoritative database. For example, the host `ns.cs.kent.edu` may run a name server for the domain `cs.kent.edu`.

A name server answers queries from resolvers and provides either definitive answers or referrals to other name servers. The DNS database is set up to handle network address, mail exchange, host configuration, and other types of queries, with some to be implemented in the future.

The ICANN and others maintain *root name servers* associated with the root node of the DNS tree. In fact, the VeriSign host `a.root-servers.net` runs a root name server. Actually, the letter `a` ranges up to `m` for a total of 13 root servers currently.

Domain name registrars, corporations, organizations, Web hosting companies, and other ISPs run name servers to associate IPs to domain names in their particular zones. All name servers on the Internet cooperate to perform domain-to-IP mappings on the fly.

DNS Resolvers

A DNS resolver is a program that sends queries to name servers and obtains replies from them. A resolver can access at least one name server and use that name server's information to answer a query directly or pursue the query using referrals to other name servers. Hence, a resolver is a client that Internet applications call to obtain desired information from the DNS, such as domain to IP mapping.

Try the **Demo:** `NSLookup` at the CT website that looks up domain/IP information from the DNS.

In addition to IP address and domain name, a host's DNS records also contain MX entries for directing the domain's email flow. DNS records can provide other useful information related to a host. Table 5.1 shows common DNS record and request types.

TABLE 5.1 DNS Record/Request Types

Type	Description
`A`	Host's IP address
`NS`	Name servers of host or domain
`CNAME`	Host's canonical name, and an alias
`PTR`	Host's domain name, IP
`HINFO`	Host information
`MX`	Mail exchanger of host or domain
`AXFR`	Request for zone transfer
`ANY`	Request for all records

CT: DECENTRALIZE

Decentralize control of a large system to make it more efficient and robust.

The concept is commonplace in the way we run governments and large corporations. Often it is best that local entities have control of their affairs and cooperate to support the mission of the enterprise.

Distributing the same tasks to several workers can also provide redundancy and flexibility, even for a small group of employees in a certain department. For example, each of the several bank tellers can take care of any routine customer service without central control. Even if a teller or two are absent on any given day, the bank continues to work. Servers in restaurants work the same way.

In the beginning, the ARPANET used a file `HOSTS.TXT` to centrally record domain-IP associations kept at Stanford Research Institute (SRI). This file was updated by SRI and transmitted to connected nodes.

As the Internet expanded, we needed to keep track of the ever-changing information of connected hosts, their IP addresses, and domain names. Further, we need to instantly process domain-to-IP mapping requests from any host/service. And what is the solution? It is decentralization. But how? Well, the problem, the widespread Internet, is the solution. The DNS distributes itself over the Internet and enables local managers to provide and update their DNS databases. Thus, the DNS is decentralized over the Internet itself, making both the DNS and the Internet more robust. How clever!

5.13 Domain Registration

Anyone can obtain a domain name, often for creating a website, or some other purpose. To obtain a domain name, you need the service of a *domain name registrar*. Most will be happy to register your new domain name for a very modest yearly fee. Once registered, the domain name is property that belongs to the *registrant*. No one else can register for that particular domain name as long as the current registrant keeps the registration in good order.

ICANN accredits commercial registrars for common TLDs, including `.com`, `.net`, `.org`, and `.info`. Additional TLDs include `.biz`, `.pro`, `.aero`, `.name`, and `.museum`. Restricted domains (for example, `.edu`, `.gov`, and `.us`) are handled by special registries, such as `net.educause.edu` (for `.edu`), `nic.gov` (for `.gov`), and `nic.us` (for `.us`). Country-code TLDs are normally handled by registries in their respective countries.

5.13.1 Accessing Domain Registration Data

The registration record of a domain name is often publicly available. The standard Internet *whois* service allows easy access to this information. You can do this on the Web at `www.internic.net/whois.html`, for example. On Linux/Unix systems, easy access to whois is provided by the **whois** command,

whois *domain_name*,

which lists the domain registration record kept at a registrar. For example,

whois `kent.edu`

produces the following information

```
Domain Name: KENT.EDU

Registrant:
   Kent State University
   500 E. Main St.
   Kent, OH 44242
   UNITED STATES

Administrative Contact:
   Philip  L Thomas
   Network & Telecomm
   Kent State University
   STH
   Kent, OH 44242
   UNITED STATES
   (330) 672-0387
   pki-admin@kent.edu
```

```
Technical Contact:

   Network Operations Center
   Kent State University
   120 Library Bldg
   Kent, OH 44242
   UNITED STATES
   (330) 672-3282
   noc@kent.edu

Name Servers:
   NS.NET.KENT.EDU           131.123.1.1
   DHCP.NET.KENT.EDU         131.123.252.2
   ADNS03.NET.KENT.EDU       128.146.94.250

Domain record activated:     19-Feb-1987
Domain record last updated: 06-Jul-2011
Domain expires:              31-Jul-2015
```

Note that the `kent.edu` domain has name servers running on three hosts as listed.

Try the **Demo:** `WhoIs` at the CT website that can retrieve desired domain registration records.

5.14 Packet Switching

Data on the Internet are sent and received in *packets*. Thus, the Internet is a packet switching network. Similar to a letter, a packet envelops a small block of data with address information so the data can be routed through intermediate nodes on the network, which is shared by all connected users. The network uses routing algorithms to efficiently forward packets to their final destinations.

Because there are multiple routes from the source to the destination host, the Internet is very reliable and can operate even if parts of the network are down. Figure 5.15 shows the structure details of an IPv4 packet. The individual parts of a packet are as follows:

- **Version** (4 bits): IP version number (4 for IPv4)
- **Ihl** (4 bits): header length in number of 32-bit words
- **Type of service** (8 bits): service quality parameters
- **Total length** (16 bits): total length of the packet, IP header plus data (payload)

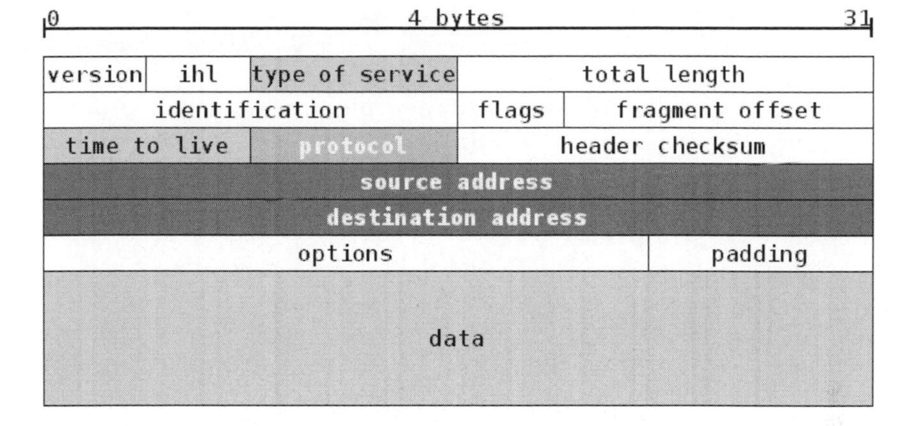

FIGURE 5.15 IPv4 Packet

- **Identification** (16 bits): an ID value assigned by packet sender
- **Flags** (3 bits): bit flags for packet fragmentation control
- **Fragment offset** (13 bits): 0-based sequence number of a fragmented packet
- **Time to live** (8 bits): number of hops allowed before the packet is dropped; zero means dropping the packet now
- **Protocol** (8 bits): network layer, for example, 1 for ICMP, 6 for TCP, and 17 for UDP
- **Header Checksum** (16 bits): checksum value used to check if the packet is received errorfree
- **Source Address** (32 bits): IPv4 address of the sender (or source) of the packet
- **Destination Address** (32 bits): IPv4 address of the receiver of the packet
- **Options** (variable length): when the value of ihl is greater than 5, options may indicate values for security, record route, time stamp, and so on.

5.15 Cloud Computing

As a result of the ever-increasing accessibility, speed, and reliability of the Internet, the computing industry has begun to offer data storage and processing power on the Internet as a service. Most are subscription based, but some can be free. The practice is known as *cloud computing*. Subscribers of cloud computing, organizations or individuals, simply enjoy the computing power without worrying about hardware/software installation, operation, maintenance, or providing user help. In fact, cloud customers don't even care where the facilities are located. They are all *in the cloud*. And the rented services are available 24x7 anywhere on the Internet from any desktop, laptop, Chromebook[2], tablet, or smartphone.

Companies, such as Amazon, Google, IBM, Oracle, Microsoft, CloudBees, Rackspace, and many others, have the economy of size to supply cost-effective cloud computing services on the Internet. For businesses large and small, cloud computing can be an alternative to owning, staffing, and operating their own IT (information technology) equipment in house. Promoters of cloud computing ask, "If you need milk, would you own a cow?" Just get the milk (the computing power you need) and let someone else worry about the cow (everything related to providing the milk).

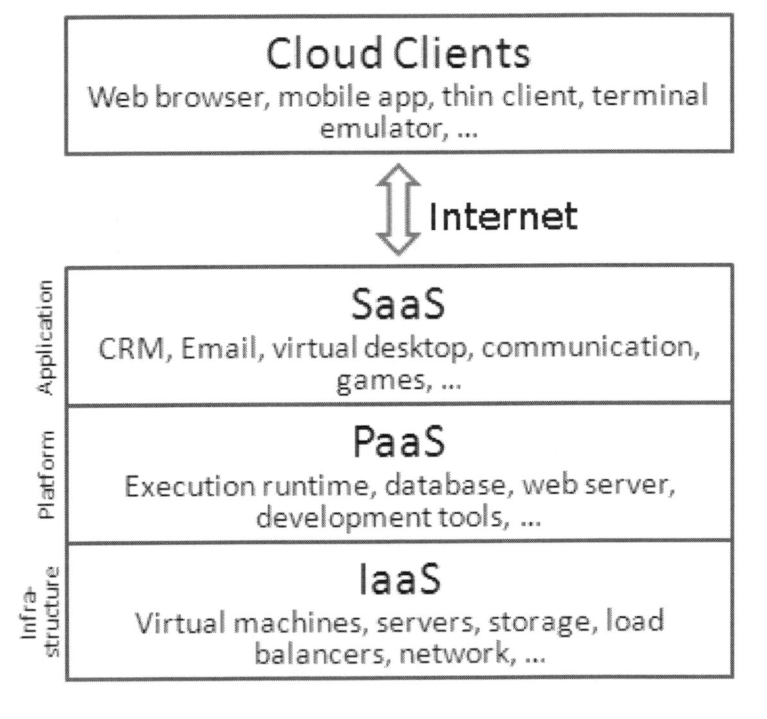

FIGURE 5.16 Cloud Computing Layers

Types of services offered by cloud service providers (CSPs) include:

- Cloud storage—Distributed, virtual, reliable, and fault tolerant storage

[2]Often comes with cloud storage.

of data that is easily accessible on the Internet/Web. Dropbox and Picasa are well-known examples. Often, cloud storage is also combined with document management.

- Software as a service (SaaS)—Software running on cloud servers accessible ondemand to subscribers, typically through a thin client via a Web browser. SaaS software performs customer relationship management (CRM), computer aided design (CAD), database management (DBM), human resource management (HRM), and many other tasks.

- Platform as a service (PaaS)—Virtual hardware–software servers, typically with OS, application programming and execution environment, database, and Web server. Customers can control and use the platform as well as develop custom applications on it.

- Infrastructure as a service (IaaS)—Virtual IT data centers complete with maintained servers, storage, and network facilities. A customer can install operating systems and develop and deploy their own applications.

Figure 5.16 shows the layered structure of cloud computing facilities.

Cloud computing has clear advantages and disadvantages, as compared to in-house solutions.

The top advantages are: less up-front investment, faster to set up and start, reducing IT personnel and equipment, more reliable and physically secure, wider access, and easier to maintain, upgrade, scale up or down.

Disadvantages are: relying on CSP for privacy and security, less in house control of IT services, much lower network speed (compared to in-house LAN), limitations of applications running remotely rather than locally.

CT: Backup in the Cloud

Backing up important data on the cloud is not hard and can save the day.

For example, modern airliners have voice and data recorders, commonly referred as "black boxes." After an accident, locating and retrieving the black box often becomes a difficult task. If critical data are continuously communicated via satellite to secure cloud storage, then it would be possible to better monitor flights in real time and to perhaps eliminate the physical black boxes.

Cloud computing is a growing industry. It makes sense for the right tasks and can be cost effective and more convenient. Individuals and IT professionals will need to weigh the pros and cons and pick the right cloud solutions. Often, a combination of in-house platforms, private cloud, and public cloud can be the best choice. Free and open cloud software, such as FOSS-Cloud (`foss-cloud.org`) and others, makes it much easier to create your own cloud services.

We presented an overview of wired/wireless LAN, WAN, and the Internet. In the next chapter, we will focus on the most important Internet application, the World Wide Web.

Meantime, the network connection is back, and you are online! Finally, you can get to your email, say hello on Skype, and hang out with your friends on Google+. Life is good and your instant messages have a new air of confidence, and appreciation. Thank goodness for the Internet.

Exercises

5.1. Is a tree also a graph? What is the difference between a tree structure and a graph structure?

5.2. On a network, what are terminal nodes? Internal nodes?

5.3. What is a router for a home LAN? What does it do?

5.4. Look into the upcoming 5G mobile network technology and compare it to 4G.

5.5. What is a networking protocol? Give three examples.

5.6. What is an IP address? A domain name? What is their relationship?

5.7. Describe the client and server model of networking.

5.8. What is packet switching? Please explain.

5.9. What is DNS, and why is it important for the Internet as a whole?

5.10. **Computize**: From your own viewpoint, list the top 3 impacts of the Internet on humankind.

5.11. **Computize**: What practices would you recommend for sending email to make it easier on the receiving end? Use sending completed homework to an instructor by email as an example to illustrate your recommendations.

5.12. **Computize**: Create a flowchart for a step-by-step procedure to submit homework by email. The procedure needs to be detailed enough so someone else (your little sister) can do it correctly by following the procedure.

5.13. **Computize**: Look into the Y2K problem. How would you apply the "indirection" idea to solve the Y2K problem?

5.14. **Computize**: Cloud computing is useful indeed. Does it inspire any CT ideas for you?

5.15. **Computize**: Look into the 9/11 terrorist attack on the USA and the Malaysia Flight 370 mystery. Then, give your views on the comment, "We know where a cellphone is all the time, how can a large commercial airliner go off course without triggering an alert in real time?"

5.16. **Group discussion topic**: *The information highway vs. the real highway.*

5.17. **Group discussion topic**: *Net neutrality.*

5.18. **Group discussion topic**: *Traffic and transportation safety and smartphones.*

Chapter 6

Home Sweet Homepage :-)

"Hello Laura, it is nice to meet you. Please (producing a business card) visit our website and find out all about me or my company." Or, in an email, "Can you believe this? See it here." Or, "Look at their menu, I am very much looking forward to our dinner together." These days, no serious professional or organization does not already have, or plans to have, a site on the World Wide Web (Web).

A website gives you your own domain name (`BigBadWolf.com`, for example). Then you'll have email addresses in the form of `someone@BigBadWolf.com`. Don't forget to add your website and email address to your business card. You'll be able to upload webpages and other information to your site for everyone to use. And that's just the beginning of the advantages.

The Web service uses the client and server model, as explained in Section 5.9. Web servers and clients communicate with the HTTP (Hypertext Transfer Protocol) and HTTPS (Secure HTTP) application-level protocols (Figure 5.9), riding on top of TCP/IP. The *Hypertext Markup Language* (HTML) is used to author webpages. A resource available on the Web is retrieved by its Web address known as a URL (*Universal Resource Locator*). For example, the CT website is at `http://computize.org`.

The Web's global impact has brought us numerous advantages, large and small. And it is transforming society at all levels as untold number of people actively blog, microblog, and otherwise express themselves with online media. It goes without saying that an understanding of the Web and how it works will be good for everyone.

6.1 What Is a Web Server?

A key factor for the Web's great success is the low cost of putting information on it. You simply find a Web hosting service to upload files for your website. Any Internet host with a good Internet connection can provide Web hosting. A Web host provides space for websites to store webpages, pictures, audio and video files, and any programs that produce Web content on the fly. More importantly, the Web host also runs a *Web server* program through which the stored websites can be accessed on the Web.

There are a number of Web servers, but Apache™, the open source Web server from `apache.org`, is the most dominant. According to *Web Technology*

Surveys (March 2015), market shares for major Web servers are Apache™ (58.4%), NGINX™ (23.3%), and Microsoft-IIS® (13.2%).

A Web server program listens to a specific networking port on the server host and follows the HTTP and HTTPS to receive requests and send responses. The standard HTTP port is 80, but can be some other designated port, such as 8080. The standard HTTPS port is 443.

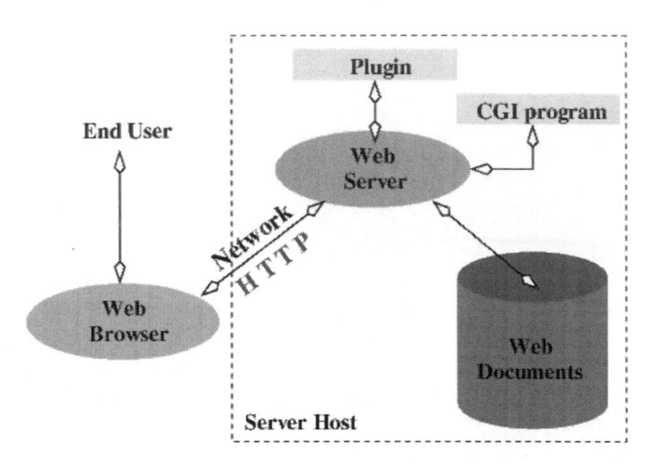

FIGURE 6.1 Web Server Overview

In response to an incoming request, a server may return a static document from files stored on the server host, or it may return a document dynamically generated by a program, such as a PHP script, indicated by the request (Figure 6.1).

A single-thread server handles one incoming request at a time, while a multithread server can handle multiple concurrent requests. A server host may have multiple copies of a Web server running to improve the handling of requests.

On a server host, webpages and other files to be accessed from the Web must be placed in the Web server's *document space* before they become available on the Web. The top folder of the document space is known as the *server root.* To further control access, files and folders inside the document space must also be given correct *access permissions* before the Web server can deliver them onto the Web.

6.2 Web Browsers

Another critical factor for the Web's success is easy-to-use clients for browsing the Web.

We access the Web from anywhere on the Internet with desktops, laptops, tablets, and smartphones. Popular Web browsers (Figure 6.2) are:

FIGURE 6.2 Top Five Web Browsers

- Vendor proprietary—Google Chrome (all platforms), Internet Explorer (Microsoft Windows), Safari (Mac OS X and iOS)
- Open source for all platforms—Firefox, Opera, and others

In terms of popularity, you have Google Chrome and Firefox enjoying 56% and 30% of the market,[1], respectively.

Browsers make surfing the Web a great experience. They keep track of your browsing history, remember and organize your bookmarks, safely keep your user IDs and passwords for different sites, cache Web data for speed, supply multiple tabs (such a convenience), and allow a high degree of user customization, including browser home page (URL loaded automatically when the browser starts), fonts, colors, helper applications and more.

Accessing the Web on the go is important and fun. Android tablets and smartphones come with Google Chrome. Apple devices come with Safari. Windows phones come with IE Mobile. Businesses also provide dedicated mobile apps to access their information and services.

6.3 A Brief History of the Web

The Web got its start in the late 1980s. In 1989, Tim Berners-Lee at the European Laboratory for Particle Physics (CERN) started to develop a suite of technologies to make the Internet truly accessible and useful to people.

- URL—The Uniform Resource Locator
- HTML—The Hypertext Markup Language
- HTTP—The Hypertext Transfer Protocol

Berners-Lee also wrote the first Web browser and server.

The simplicity of HTML makes it easy to learn and publish webpages. It caught on. In 1992–1993, a group at NCSA (National Center for Supercomputing Applications, US) developed the Mosaic visual/graphical browser (Figure 6.3). Mosaic added support for images, nested lists, and fill-in forms, and it fueled the explosive growth of the Web. Several people from the Mosaic project helped start Netscape (Figure 6.3) in 1994. At the same time, the W3

[1] 2014 statistics by w3cschool.

FIGURE 6.3 Mosaic and Netscape

Consortium (W3C) was formed and housed at MIT as an industry-supported organization for the standardization and development of the Web.

6.4 URLs

The Web uses *Uniform Resource Locators* (URLs) to identify (locate) many kinds of resources available on the Internet. URLs are used by Web browsers to request and retrieve information. We know URLs can locate webpages. But they can also identify pictures/images and audio and video media, as well as Internet/Web services.

A URL usually has the form

scheme`://`*serverhost*`:`*port*`/`*pathname*`?`*query_string*

that consists of several parts. Let's break the URL down.

- The *scheme* part indicates the information service type and therefore the protocol. Common schemes include `http` (Web service), `ftp` (file transfer service), `file` (local file system), `https` (secure Web service), and `sftp` (secure file transfer service). Many other schemes can be found at `www.w3.org/addressing/schemes`. Because `http` is the default scheme, it, together with the trailing `://` can sometimes be omitted.
- The *serverhost* part, separated from the *scheme* by `://`, is a domain name or IP address that identifies a server host.
- The *port* number is optional. It is a number given with a : prefix to identify a networking channel on the host. Standard Internet services are assigned default ports (for example, 80 for `HTTP` and 443 for `HTTPS`). The port number is needed only if the server program does not use its default port.
- The *pathname* part is optional. If omitted, the URL brings up the *homepage* (site entry) of a website. The pathname is a filename, with a `/` prefix, relative to the server root folder on the server host. If this pathname has a trailing `/` character, it represents a directory rather than a data file.

- The *query string* part is optional. When the pathname leads to an executable program that dynamically produces an HTML or other valid file to return (Section 6.8), a query string, given after a `?`, can provide input values to that program. A query string takes the form of *name*=*value* pairs separated by `&`.

Here are some examples,

`http://www.kent.edu`	(Ket State U. homepage)
`http://w3.org/`	(W3C site)
`https://chase.com`	(Chase Bank secure site)
`http://computize.org/example.html`	(CT site example page)
`https://amazon.com/.../home?ie=UTF8`	(`amazon.com` after login)
`ftp://webtong.com`	(Public FTP `webtong.com`)
`file:///C:/Users/pwang/Desktop/a.jpg`	(Picture on local Desktop)

URLs are critical in Web operations. You can enter any valid URL into the `Location` box of any Web browser to reach the target resource. When a URL specifies a directory, a Web server usually returns an *index file*, typically named `index.html`, for that directory. Otherwise, it may return a list of the filenames in that directory. Thus, for example, the URL `http://cnn.com` is the same as `http://cnn.com/index.html`.

URLs are also used in webpages to link to other webpages and resources, inside or outside a particular website. The cross-links among webpages globally form a worldwide web structure. Because of its importance, many applications, including email readers, PDF readers, text/document editors, presentation tools, and shell windows, recognize the `http` URL and, when you click on it, will launch the default Web browser.

Relative URLs

Within an HTML document, you can link to another document served by the same Web server by giving only the *pathname* part of the URL. Such links are examples of *relative URLs.*

- A relative URL with a leading `/` (for example, `/file_xyz.html`) refers to a file under the *server root*, the top-level directory controlled by the Web server.

- A relative URL without a leading `/` points to a file relative to the location of the document that contains the URL in question. Thus, a simple `file_abc.html` refers to that file in the same directory as the current document.

CT: BE AWARE OF THE IMPLICIT CONTEXT

Be aware of the implicit context. It can bring convenience and efficiency or confusion and misunderstanding.

Implicit contexts are everywhere and happen all the time. When dialing a local phone number we may skip the country code and sometimes even the area code. When addressing a domestic letter, we do not have to indicate the country. When mentioning an address to a friend, we save our breath on the country, state, or even city. On the Internet, inside the domain `cs.kent.edu`, we can refer to the host `tiger.cs.kent.edu` as simply `tiger`.

When building a website, it is advisable to code webpages using URLs relative to the current page as much as possible. This makes it easy to reorganize the file/folder structures of a website and to move the entire website to another location on the local file system or to a different server host.

When communicating, the parties must use the "same implicit context" or misunderstandings can happen. For example, "You should be honest" can be stating a principle (editorial "you") or an accusation. A day of the week is relative to the week. Therefore, people must pay attention to the sent date and time when reading email or text messages (CT: PAY ATTENTION TO DETAILS, Section 4.6.1). But why depend on others being careful? Instead of terms such as "Saturday," "yesterday," "tomorrow," or "next week", we should always state a specific date and time in our messages.

In sly advertisements, clever manipulation of the implicit context is often used to mislead. No wonder why legal documents are so lengthy, repetitive, and formal.

6.4.1 URL Encoding

According to the URL specification (RFC1738), only the following characters may be included directly within a URL.

- Reserved characters: `! * ' ( ) ; : @ & = + $ , / ? # [ ]` used for their reserved purposes (supporting the URL syntax)
- Unreserved characters: `0`–`9`, `a`–`z`, `A`–`Z`, and `$ - _ . ~`

Other characters (such as SPACE, NEWLINE, `\`, `"`, and so on), reserved characters not used for URL syntax, and non-ASCII (Section 2.6.1) UNICODE characters (Chinese characters, for example) may cause problems (unsafe) if used directly in a URL. To include such a character, it should be encoded following *percent encoding* rules.

- To percent encode an unsafe ASCII character, replace it with a three-character sequence `%hh` (*percent encoding*), where *hh* is the character's byte code in hexadecimal. For example, `~` is `%7E` and SPACE is `%20`. Thus, a relative URL to the file "`chapter one.html`" becomes

  ```
  chapter%20one.html
  ```

  and a relative URL to the strangely-named document file `Indeed?.pdf` becomes

  ```
  Indeed%3F.pdf
  ```

  It is clear why. The `?` character unencoded would make `.pdf` a query string, not part of the file name.

- To percent encode a non-ASCII character, UTF-8 encode (Section 2.6.2) the character into two or more bytes, then percent encode each byte. For example

  ```
  U+738B    王    %E7%8E%8B
  U+58eb    士    %E5%A3%AB
  U+5f18    弘    %E5%BC%98
  ```

See **Demo:** `PercentEncode` on the CT website for an interactive tool.

CT: Wear Different Hats

Realize and provide a clear indication when the same entity is to perform a different function.

It is not unusual for a person to have multiple roles to play. For example, a policeman can be on or off duty. A door can be an entrance or an exit. A road can be oneway or twoway. Often, the distinction is important to avoid confusion, and we use different hats, labels, signs, or uniforms as indications. In case of the one-way street, the sign may mean life or death.

In computing, characters often must do multiple duties. This is simply because there are not enough characters on the keyboard to satisfy all the varied needs in different situations. For example, in many programming languages, character strings are enclosed in double quotes (`"`). But that begs the question: "What if a double quote is part of a string?" The problem is caused by `"` performing double duty as delimiters of strings and as just a character. And the solution? Place a \ in front of `"` to *escape* it from being treated as terminating a string. The JavaScript code

```
str_a = "The double quote (\") character.";
```

is an example. Now, the BACKSLASH (\) doubles as an *escape character*, itself must then be escaped in a string:

```
str_b = "The backslash (\\) character.";
```

The escape character performs the same function as a hat or a label for the next character in a string. The URL percent encoding is basically the same story—the `%` character escapes a 2-character sequence that represents an arbitrary byte. And, again, the character `%` must itself be percent encoded (`%25`) to be part of a URL.

6.5 HTML and HTML5

HTML (the Hypertext Markup Language) is used to structure webpage contents for easy handling by Web clients on the receiving end. From its simple start in 1989, HTML has been constantly evolving and maturing. Beginning with HTML 4.0, the language has become standardized under the auspices of the W3C (World Wide Web Consortium), the industry wide open standards organization for the Web. Subsequently, by making HTML 4.0 compatible with XML (eXtensible Markup Language Section 9.7.1), XHTML became a widely used new standard. Today, the Web is moving toward HTML5, the next-generation HTML standard, which brings many new features and APIs (application programming interfaces). HTML5 makes it easier to provide dynamic user interactions and promises to transform the Web into an even more useful and powerful tool.

A document written in HTML contains ordinary text interspersed with *markup tags* and uses the `.html` filename extension. The tags mark portions of the page as heading, section, paragraph, quotation, image, audio, video, link, and so on. Thus, an HTML file consists of two kinds of information: contents and HTML tags. The HTML code provide webpage organization and structure information to make automatic processing of the contents easier. An HTML tag takes the form <*tag*>. A *begin tag* such as `<h1>` (level-one section header) is paired with an *end tag*, `</h1>` in this case, to mark content in between. Table 6.1 lists some frequently used tags.

The following is a sample HTML5 page (**Demo:** `Sports`):

```
<!doctype html>
<html xmlns="http://www.w3.org/1999/xhtml"
      lang="en" xml:lang="en">
<meta charset="UTF-8"/>
<head> <title>A Basic Webpage</title> </head>
<body><section>
   <h1>Big on Sports</h1>
   <p>Sports are fun and good for you ...</p>
```

TABLE 6.1 Some HTML Tags

Meaning	HTML Tag	Meaning	HTML Tag
Entire page	`<html>...</html>`	Paragraph	`<p>...</p>`
Meta data	`<head>...</head>`	Unnumbered list	`<ul>...</ul>`
Page title	`<title>...</title>`	Numbered list	`<ol>...</ol>`
Page content	`<body>...</body>`	List item	`<li>...</li>`
Level *n* heading	<h*n*>...</h*n*>	Comment	`<!--...-->`

```
   <p> What is your favorite sport? ...
   And here is a short list: </p>
   <ol>
     <li> Baseball </li> <li> Basketball </li>
     <li> Tennis </li> <li> Soccer </li>
   </ol>
</section></body></html>
```

Figure 6.4 shows the Big on Sports page displayed by Firefox.

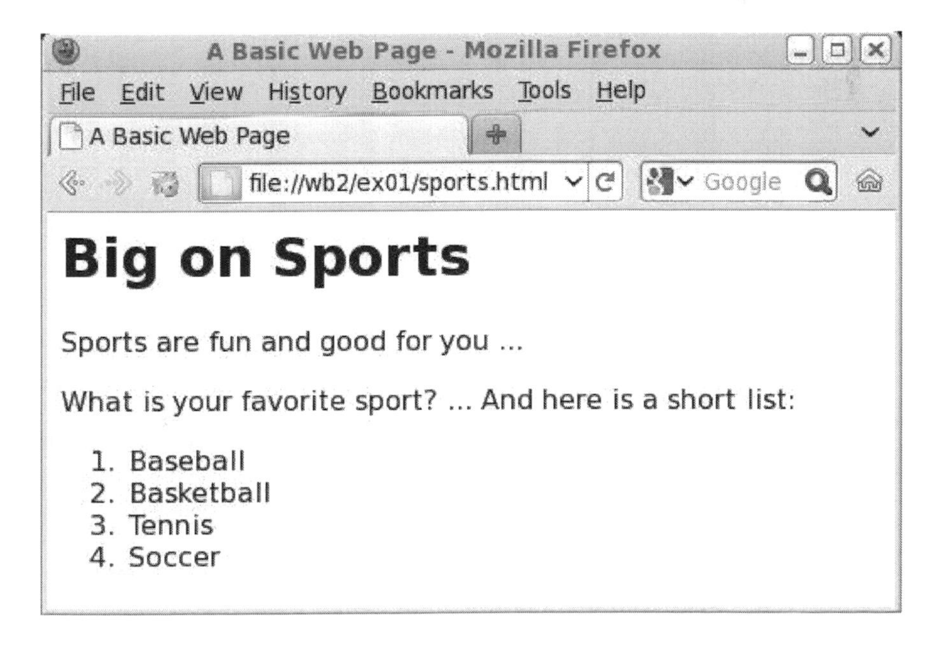

FIGURE 6.4 A Sample Webpage

CT: Mark It Up

Organize information in a document by identifying and delimiting its parts. Marked-up documents are easier to use, exchange, and process mechanically.

The idea is hardly new or surprising. Take this textbook, for example; we organized it into chapters, sections, and subsections. It has a table of contents, and an index, among other things. The organization is achieved through headers, page formats, and other visual conventions.

For textual documents, such as webpages, markup elements or tags are used to indicate the start and end of parts, such as headings, paragraphs, tables, images, quotations, links, and so on. A marked-up document can be easily transmitted and processed by applications on receiving host computers.

6.6 Webpage Styling

While HTML takes care of page structure, the way information is actually presented (visually or otherwise) to the end user is controlled by the Web browser, user-defined styling preferences, and *styling rules* that come with the webpage.

Styling rules are coded in *Cascading style sheets* (CSS) and attached to different parts of a webpage. Style rules are usually placed in files separate from the webpage. Isolating page styling from page structure makes it easy for Web designers to reuse styling rules in different pages and to enforce consistent visual styling over an entire website.

For example, if we want to make all level-one headers dark blue, we can use this CSS rule:

```
h1 { color: darkblue }
```

Thus, HTML makes webpages easy to read by programs, while CSS makes them easy to read by humans.

CSS has also evolved through the years to provide more features and functions for various styling needs. The current standard is CSS3. Experiment with HTML and CSS with the **Demo:** `CodeTester` at the CT site.

6.7 Web Hosting

Web hosting is a service for individuals and organizations to place their websites on the Web. Hence, publishing on the Web involves:

1. Designing and constructing the pages and writing the programs for a website

2. Placing the completed site with a hosting service

Colleges and universities host personal and educational sites for students and faculty without charge. Web hosting companies provide the service for a fee.

Commercial Web hosting can provide secure data centers (buildings), fast and reliable Internet connections, specially tuned Web hosting computers, server programs and utilities, network and system security, regularly scheduled backup, and technical support. Each hosting account provides an amount of disk space, a monthly network traffic allowance, email accounts, Web-based site management and maintenance tools, and other access, such as FTP and SSH/SFTP.

> **CT: Really Use Your Website**
>
> *Put your website to serious use. Manage it appropriately. Update it diligently. Make it an integral part of your organization.*

A website is far more than a static online advertisement. It is a window to the world. Take advantage of all that the Web can do to make your organization more effective and efficient. Integrate the access, modification, and management of your site into your business operations. Always make sure information on the site is up-to-date.

To host a site under a given domain name, a hosting service associates that domain name to an IP number assigned to the hosted site. The domain-to-IP association is made through DNS servers and Web server configurations managed by the hosting service.

> **CT: Be Careful with Online Information**
>
> *Be critical; don't believe everything online. Avoid spreading untruth.*

Easy online sharing is a powerful and positive force in the digital age. With all kinds of information available on the Web, Internet, and by email, we must also be keenly aware of an unpleasant fact, that not all such information is accurate or even true. The good news is, by digging a little deeper (a few Web searches, for example), you can usually find out. Too many have unwittingly sent onward to friends rumors, falsehoods, or baseless claims, even causing bad information to *go viral* sometimes. Let's not participate in such silliness.

6.8 Dynamic Generation of Webpages

Webpages are usually prepared and set in advance to supply some predetermined content. These fixed pages are *static.* A Web server can also deliver *dynamic pages* that are generated on the fly by programming on the server side. Dynamic pages bring many advantages, including:

- Managing user login and controlling interactive sessions
- Customizing a document depending on when, from where, by whom, and with what program it is retrieved
- Collecting user input (with HTML forms), processing such input data, and providing responses to the incoming information
- Retrieving and updating information in databases from the Web
- Directing incoming requests to appropriate pages; redirection to mobile sites is an example
- Enforcing certain policies for outgoing documents

Dynamic webpages are not magic. Instead of retrieving a fixed file, a Web server calls another program to compute the document to be returned and perhaps perform other functions. As you may have guessed, not every program can be used by a Web server in this manner.

A Web server invokes a server-side program by calling it and passing arguments to it and receiving the results thus generated. Such a program must conform to the Common Gateway Interface (CGI) specifications governing how the Web server and the invoked program interact (Figure 6.5).

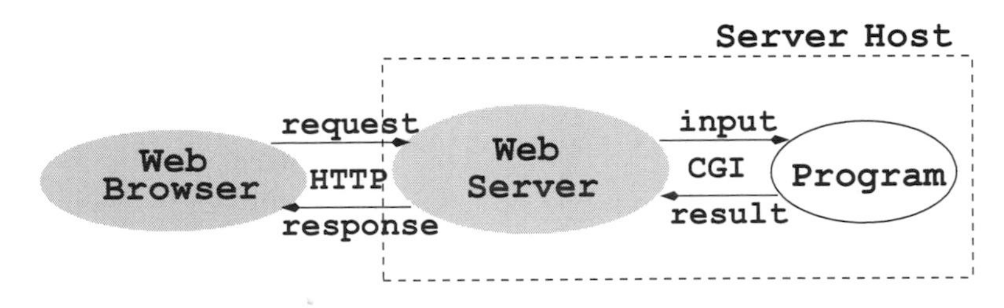

FIGURE 6.5 Common Gateway Interface

A CGI program can be written in any programming language as long as it follows the CGI specification and can be invoked by the Web server. The Web server and a CGI program may run as independent processes and interact through interprocess communication. Or, the external program can be loaded into the server and run as a plug-in module.

6.8.1 Active Server Pages

The dynamic generation of pages is made simpler and more integrated with webpage design and structure by allowing a webpage to contain *active parts* (Figure 6.6) that are treated by the Web server and transformed into desired content on the fly as the page is retrieved and returned to a client browser.

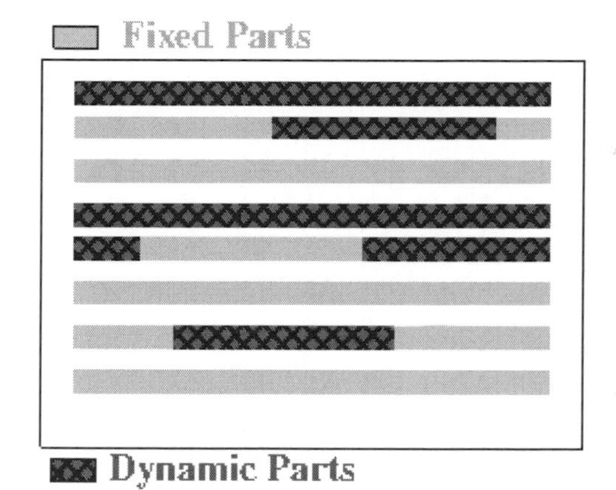

FIGURE 6.6 An Active Page

The dynamic (active) parts in a page are written in some kind of notation to distinguish them from the fixed parts of a page. The ASP (Active Server Pages), JSP (Java Server Pages), and the popular PHP (Hypertext Preprocessor) are examples. With PHP, the active parts are enclosed inside the bracket `<?php ... ?>` and embedded directly in an HTML page or other types of Web document. For example, inside an active page, code such as

```
<p>Today's date is:  <?php echo(date("l M. d, Y")); ?><p>
```

may appear. The date is dynamically computed and inserted in the HTML paragraph. Here is a result line the code would generate:

```
Today's date is: Wednesday Sep. 2, 2015
```

When active pages are treated by modules loaded into the Web server, the processing is faster and more efficient, compared to external CGI programs. PHP usually runs as an Apache module and can provide excellent server-side programming and support.

6.8.2 Database Access

Dynamic webpages are often generated from information stored in databases (Section 9.10). A database is an efficiently organized collection of data for a specific purpose. Database systems use the standard SQL (Structured Query Language; Section 9.10.2) for access and update of information in databases (Figure 6.7).

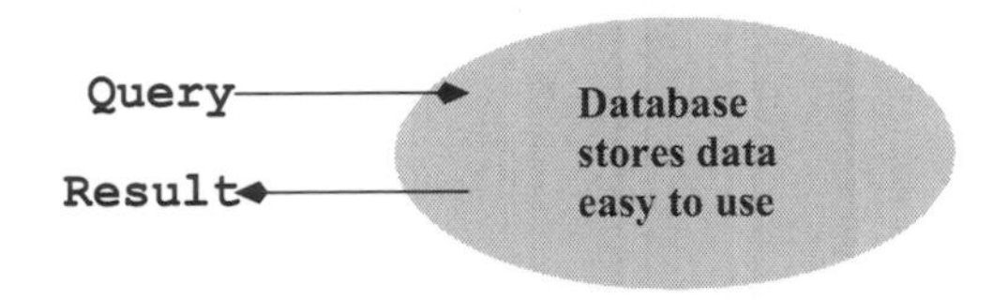

FIGURE 6.7 Database Function

Examples of databases abound: employee database, membership database, customer account database, airline (hotel) reservation database, user feedback database, inventory database, supplier or subcontractor database, and so on.

Relational database systems (RDBMS; Section 9.10.1) support the management and concurrent access of *relational databases.* A relational database is one that uses tables to store, organize, and retrieve data.

In today's fast-moving world, online access to databases is increasingly important for businesses and organizations. Using a Web interface to provide such online access has become the norm. In addition to providing access to databases, many websites also employ databases for their own purposes, such as user accounts, product inventory, blogging and forum support, just to name a few.

6.9 Client-Side Scripting

Modern browsers make the Web useful for everyone by providing a convenient user interface that usually supports keyboard, mouse, and touch screen interactions as well as video and audio presentations.

The actions of a Web browser can be defined and controlled by programming within a webpage. Such programming can supply customized user experiences and make webpages more responsive and useful for end users. The programs execute within the browser, which runs on the client host, the computer used to access the Web. For all major browsers, *JavaScript* is the standardized scripting language for client-side programming. Because the JavaScript language standard has been developed and maintained by the *ECMA* (European Computer Manufacturer Association), the language is also known as *ECMAScript* (`ecma-international.org`).

With JavaScript, a webpage can define reactions to user interface events (Section 4.6.2), verify correctness/completeness of user input, exchange information with the page's server while displaying a page, change/update and otherwise manipulate the page display, and much more. Because JavaScript runs on the client, it takes advantage of the processing power of the client host and can potentially lessen the load on the Web server.

6.10 Hypertext Transfer Protocol

As we have mentioned, Web browsers and Web servers communicate following HTTP, the *Hypertext Transfer Protocol*. It does not matter which browser is contacting what server; as long as both sides use the same protocol, everything will work.

In the early 1990s, HTTP gave the Web its start. HTTP/1.0 was standardized in the first part of 1996. Important improvements and new features have been introduced in HTTP/1.1, and it is now the stable version.

HTTP is an application layer (Figure 5.9) protocol that sits on top of TCP/IP, which provides reliable two-way connection between the Web client and Web server. We don't need all the details to understand the basics of HTTP.

1. A Web client, usually a browser but can be any user agent (UA), sends an HTTP *query* to a server.

2. A Web server, upon receiving a query, sends back an HTTP *response*.

A query and a response form an HTTP *transaction*. Each transaction stands alone and has no protocol-provided means to be correlated with any other transaction. Figure 6.8 illustrates an HTTP transaction.

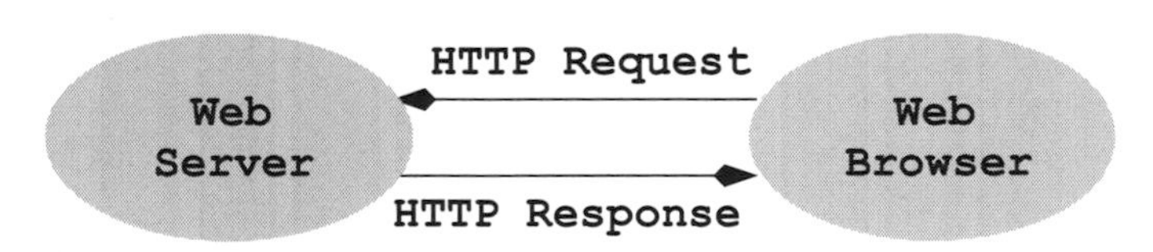

FIGURE 6.8 An HTTP Transaction

A simple HTTP transaction goes as follows:

1. *Connection*—A browser (client) opens a connection to a server.

2. *Query*—The client requests a resource controlled by the server.

3. *Processing*—The server receives and processes the request.

4. *Response*—The server sends the requested resource, or an error, back to the client.

5. *Termination*—The transaction is finished, and the connection is closed unless it is kept open for another request immediately from the client on the other end of the connection.

HTTP governs the format of the query and response messages (Figure 6.9). Basically, each query or request consists of an *initial line*, one or more *header lines* and an optional *body*. The initial line and header lines are textual (ISO-

initial line	(different for query and response)
HeaderKey1: value1	(zero or more header fields)
HeaderKey2: value2	
	(an empty line with no characters)

Optional message body contains query or response data.
Its data type and size are given in the headers.

FIGURE 6.9 HTTP Query and Response Formats

8859-1). Each line should end in RETURN and NEWLINE, but it may end in just NEWLINE.

The initial line identifies the message as a query or a response.

- A query line has three parts separated by spaces: a *query method* name, a local path of the requested resource, and an HTTP version number. For example,

```
GET   /path/to/file/index.html   HTTP/1.1
HOST: domain_name
```

  or

```
POST   /path/script.php   HTTP/1.1
HOST: domain_name
```

  The `GET` method requests the specified resource and does not allow a message body. A `GET` method can invoke a server-side program by specifying the CGI or active-page path, a question mark, and then a *query string*:

```
GET /CT_join.php?name=value1&email=value2   HTTP/1.1
HOST:  computize.org
```

  Unlike `GET`, the `POST` method allows a message body and is designed to work with HTML forms for collecting input from Web users. The `POST` message body transmits name-value pairs just like a query string.

- A response (or status) line also has three parts separated by spaces: an HTTP version number, a status code, and a textual description of the status. Typical status lines are

  ```
  HTTP/1.1   200   OK
  ```

  for a successful query or

  ```
  HTTP/1.1   404   Not Found
  ```

  when the requested resource cannot be found.

- When the HTTP response contains a message body, the `Content-Type` and `Content-Length` headers are set so the client will know how to process it.

The Web borrowed the content type designations from the Internet email system and uses the same MIME (Multipurpose Internet Mail Extensions) defined content types. Hundreds of standard MIME content types are listed at the IANA site (`iana.org/assignments/media-types/media-types.xhtml`).

The content type information allows browsers to decide how to process the incoming content. HTML, text, images, audio, and video may be handled by the browser directly. Other types, such as PDF and Flash, are usually handled by plug-ins or external helper programs.

When using a browser to access the Web, the HTTP messages between it and the Web servers are kept behind the scenes. But it is possible to expose these messages and gain real experience with HTTP. See **Demo:** `Http` at the CT website.

While HTTP transmits information in the open, HTTPS (HTTP Secure) is a secure protocol that simply applies HTTP over a secure transport layer protocol *Transport Layer Security* (TLS 1.2) that is derived from the earlier *Secure Sockets Layer* (SSL). See Section 7.2 for more information on how this security feature works.

6.10.1 HTTP Caching

An important improvement of HTTP 1.1 over HTTP 1.0 is the introduction of caching for HTTP responses. On the Web, a great deal of contents are not changing often with time. These include static webpages, images, graphics, styling code, scripts, and so on. Saving a copy of such data can avoid a lot of unnecessary work of requesting and retrieving the same data over and over again from *origin servers.* Browsers (user agents) and caching proxy servers are able to serve data from their cache when they know or can verify that the data are still current and unchanged on the servers where they originated.

A caching proxy server accelerates requests by providing contents from its cache. Caching proxies keep local copies of popular resources so large organizations can greatly reduces their Internet usage and costs, while significantly enhance performance. Most ISPs and large businesses employ caching proxies.

The HTTP caching scheme significantly cuts down round-trip Web traffic to origin servers and reduce response time to users. This explains why it is slower the first time you visit a website. Then it is lightning fast when you visit again.

> **CT: CACHE FOR SPEED**
>
> *Use cache to increase efficiency and speed. In many situations, significant improvement may result from storing the right items in a cache.*

Take phone numbers for example. You remember certain frequently used numbers in your head (the cache) but have to consult your phone list, or even the white pages for other numbers. You have your favorite pots and pans handy in the kitchen and many others stored in the basement. You have important items in your wallet/purse, and many other things you don't carry with you. And we already know that computer memory is organized into on-chip cache, RAM, and hard disk, a multilevel caching scheme.

Toward the end of 2013, when a team of super coders were helping to rescue and fix the `healthcare.gov` site, one immediate technique they used was introducing a *database cache* so that frequent queries could be separated from other queries into the huge database. The database cache reduced congestion, and they were able to lower the average page access time from 8 seconds to about 2 seconds. Later, with continued improvements, the access time was reduced to below 0.35 seconds. The rescue work may well have saved the Affordable Care Act from disaster.

6.11 Website Development

The simplicity of HTML makes it seem deceptively easy to create websites. Well, that may be true for very simple webpages with basic information. But, an inviting, attractive, and effective website requires much more effort and expertise.

A website is often a combination of online advertisement, product and service information, sales, shipping, and customer service, as well as other business functions, such as recruiting and investor relations.

CT: DEVELOP FOR USERS

> *Walk in the shoes of users of your product. Let user-centered thinking guide product design and development.*

Make sure your users will not say: "These people seem to have never used their own product!"

For Web development, in addition to carefully preparing the text, image, and multimedia contents, a well-designed website must:

- Identify the target audiences and anticipate their needs
- Have a logical and attractive user interface design
- Make finding information and doing business on the site fast and easy
- Ensure a pleasant and rewarding browsing experience
- Provide for easy update, revision, and management

To achieve these goals, it takes expertise in usability, visual communication design, site architecture and navigation, and copy editing, as well as programming techniques. And it takes time and effort to test, debug, and deploy. In other words, great websites need professional help and cooperation on the part of the site owner. Sometimes, even development professionals can mess up. The infamous `healthcare.gov` website launch debacle in October 2013 is an example.

6.12 Web Search Engines

The Web is so easy to access and contains so much information that the answer to almost any question is just a Web search away.

Many *search engines* are available, but Google remains predominant (Figure 6.10; source: `netmarketshare.com`).

A search engine makes finding information on the Web easy by working hard to gather data about what's where online and organize the collected data into indexes for efficient search. Because data online change constantly, the job of a search engine is never done. It must roam the Web continuously to update its indexes. The equipment, algorithms, and the exact ways a search engine works are closely guarded secrets.

Generally, search engines use automated *robots* and/or manual submissions to collect indexing information. A robot or *crawler* is software that visits webpages and follows links in them to recursively visit connected pages. Meta

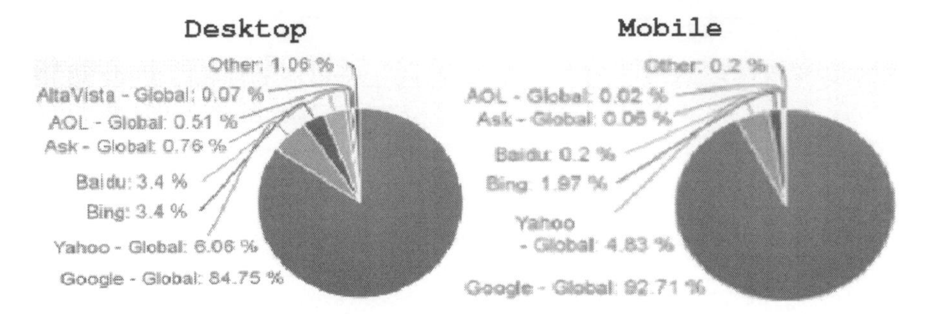

FIGURE 6.10 Search Engine Market Shares (05/2015)

information about visited pages, together with manually submitted data, are organized and deposited in databases. When a user enters a search request, the search is conducted in the databases, not on the open Web.

> **CT: Google It**
>
> *The answer is "Google."*

Doing research, curious about something, or just playing trivia? Don't bother, just Google it first.

When doing a search, try to be precise about what you are looking for. Start with the information type such as "sports," "science," "finance," "health," "entertainment," "politics," and so on. If you are looking for something local, be sure to add the location such as "Ohio" or "Kent, Ohio." Follow that with specific keywords for the search. Often, you'll get what you want in the first try. If not, you can refine your search accordingly. Getting desired information immediately online and determining the reliability of such information is a skill everyone needs to develop.

> **CT: Believe It or Not**
>
> *The Internet and Web are open. Therefore, not all information is correct or accurate.*

So, don't believe everything you see or read on the net. Use common sense. Double check. You'll soon get the correct information.

6.13 Web Services

Most people think of the Web as a vast collection of webpages ready to be visited. That is certainly true. But the Web is more than that. It is also widely used to make computing powers available to remote clients via HTTP. Such computational services are known as *Web services.*

A Web service is a resource or program on the Web that can be invoked with an HTTP request. A Web service usually computes a result based on the request input and sends back a result in a well-defined format. The most widely used Web service result formats are XML and JSON (*JavaScript Object Notation*). A Web service is like a remote procedure, it runs on a remote computer and provides specific results useful for clients. But, unlike remote procedures, Web services will always use HTTP for request and response (Figure 6.11).

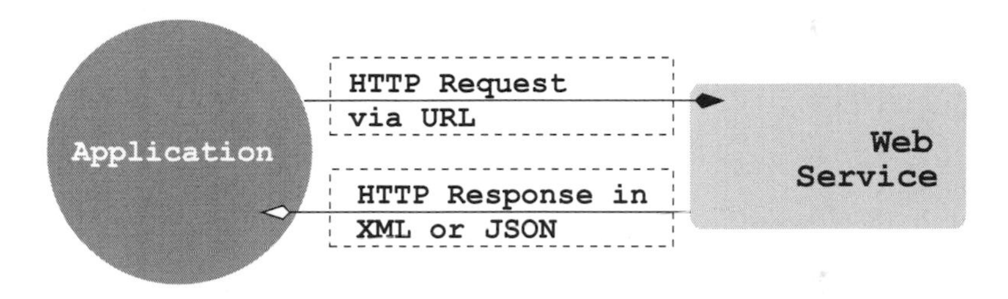

FIGURE 6.11 Web Service Overview

In the early days of Web services, the SOAP (Simple Object Access Protocol) was widely used. Contemporary Web service and client developers prefer REST[2] (*REpresentational State Transfer*), which is an academic term for URLs, query strings, and request/response bodies used in HTTP. In other words, REST-based Web services receive HTTP POST/GET requests and send back results, often in XML or JSON, in an HTTP response body.

Application programs running on a computer connected to the Internet can easily make requests to Web services and obtain results to be used in the application. For example, a news reader application can obtain news feeds (the simplest kind of Web service) via HTTP requests and then use the results (RSS[3] documents in XML) for displaying and retrieving news articles. On a Web host, we can use a server-side script, in PHP for example, to dynamically obtain information from Web services. Such server-side scripts can also be invoked by webpages, originating from the same server, via AJAX from the client side[4]. This way, a webpage can access indirectly any Web services available.

---

[2]See CT: Capture the State, Section 4.9.

[3]Rich Site Summary

[4]For security reasons, browsers only allow JavaScript to network back to its originating host.

Example Web Services

There are many Web services throughout the world. Here are some examples:

- Amazon Web services—for Amazon vendors and for providing cloud computing.
- Google Maps API Web services—for requesting Maps API data to be used in your own Maps applications; available to registered users.
- PayPal Web services—for automating various aspects of PayPal payment processing; available for PayPal account holders.
- National Oceanic and Atmospheric Administration's National Weather Service Web service—for hourly updated weather forecasts, watches, warnings, and advisories; available to the public.
- FAA airport status Web service—for US airport status, delays, and airport weather; available to the public. See **Demo: `FAA`** on the CT site.
- United Parcel Service (UPS) Web services—for address verification, shipping rates, tracking, and other services; available for registered customers. See **Demo: `Address`** on the CT site.

Online listings and directories for Web services are available. The *Web Services Directory* at `programmableweb.com` is an example.

6.14 Standard Web Technologies

In summary, a number of technologies enable the Web to work as it does. These include networking protocols, data encoding formats, clients (browsers), servers, webpage markup and styling languages, and client-side and server-side programming.

The Web can deliver text, images, animation, audio, video, and other multimedia content. Standard and proprietary media formats, tools, and players are also part of the Web. The World Wide Web Consortium (W3C) is a nonprofit organization leading the way in developing open Web standards.

Core Web technologies recommended by the W3C include:

- HTTP/HTTPS—The Hypertext Transfer Protocol employed by the Web. The current version is HTTP 1.1.
- HTML5—The new standard markup language, and its associated technologies, for coding regular and mobile webpages.
- CSS3—The current Cascading Style Sheet standard, offering improved styling, transformation, animation, and styling subject to media conditions (media queries).

- JavaScript—A standard scripting language for browser control and user interaction.
- DOM—Document Object Model, an application programming interface (API) for accessing and manipulating in-memory webpage style and content.
- PHP—A widely used server-side active page programming language and tool that is open and free.
- MySQL—A relational database system that is freely available and used widely for online business and commerce.
- DHTML—Dynamic HTML, a technique for producing responsive and interactive webpages through client-side programming.
- SVG, MathML, and XML—Scalable Vector Graphics (for 2D graphics; Section 9.2.3) and Mathematics Markup Language are part of HTML5. They are important applications of XML, the Extensible Markup Language (Section 9.7.1).
- AJAX—Asynchronous JavaScript and XML providing client-side JavaScript-controlled access to the Web and Web services.
- Web Services—Combining HTTP and XML or JSON to serve data on the Web to other programs.
- HTML5-related APIs—JavaScript-based APIs introduced by HTML5 and other projects for a good number of useful purposes.
- LAMP—Industry standard Web hosts supported by Linux (OS), Apache (Web server), MySQL (database), and PHP (active page).

The items listed here represent open/industry standards and best practices that are not private or proprietary.

With the modern Web, you can get up-to-the-moment news, find answers to any questions, satisfy any curiosity, and air your views on forums, blogs, and social media. So, fire up your browser, get online, and enjoy the Web. Life is sweet!

Exercises

6.1. Explain the Web as an Internet service in terms of client, server, and protocols.

6.2. Give your own reasons why you think the Web is so successful.

6.3. What is universal about URLs?

6.4. How is the scheme of a URL related to the port number? Please explain.

6.5. Find the percent encoding for 孔子, the Chinese name of Confucius.

6.6. Follow the address markup in Section 6.5 and invent a markup of your own.

6.7. What is client-side scripting? What standard language is used on modern browsers?

6.8. What is a database? How is the Web related to databases? Please explain.

6.9. HTTP is said to be *stateless*. Please explain.

6.10. Explain how Web search engines work and why we say, "The answer is Google."

6.11. **Computize**: Give examples of your own of an entity doing double or multiple duties and the indications used to distinguish among its roles. (Hint: HTML)

6.12. **Computize**: You and your boss are working hard on a project, burning the midnight oil. Your boss said to you, "Let's call it a day, but we need to finish this by the end of tomorrow."

6.13. **Computize**: Critically discuss the notions "scroll up", "scroll down" on Web browsers.

6.14. **Computize**: Name the ways you can personally contribute to the Web and make the world a little better.

6.15. **Group discussion topic**: *Use of terms such as he, she, it, tomorrow, yesterday, and 9 o'clock in communication.*

6.16. **Group discussion topic**: *How things were done before the Web.*

6.17. **Group discussion topic**: *I want my own website.*

6.18. **Group discussion topic**: *The impact of the Web on newspapers and magazines.*

6.19. **Group discussion topic**: *The impact of the Web on teaching and learning.*

6.20. **Group discussion topic**: *The impact of the Web on science and technology.*

6.21. **Group discussion topic**: *The impact of the Web on government.*

6.22. **Group discussion topic**: *The impact of the Web on businesses.*

6.23. **Group discussion topic**: *The impact of the Web on individuals.*

Chapter 7

Keeping It Safe

"Oh, not again, I can't remember my password!" Has this ever happened to you? I guess the answer is "yes," and not just once. Actually, before the password question even comes up, you need another thing, namely, your user ID. Each login requires a user ID and a password. But why is login required anyway? Life would be simpler without it.

Online merchants protect your account by making sure only you can access it. The user ID part of login is for *identification*, indicating who you are. The password part is for *authentication*, verifying who you are. Thus, it is very important that you don't share your login information with anyone unless you trust that person to carry out business for you online.

Cybersecurity is a vital concern for users, developers, and service providers of computers and networks. Among security concerns, identity verification, or *authentication*, is the most fundamental. Here is a list of basic concepts.

- *Confidentiality*—Keeping information secret and guarding against unwanted exposure or access
- *Integrity*—Preserving the original state of information, making sure it is unchanged and unmodified and any modifications are easily detectable
- *Authentication*—Verifying the the person, process, or system claiming to have a certain identity
- *Authorization*—Granting a set of correct permissions to an authenticated party
- *Availability*—Making sure that systems, services, or data remain accessible and usable
- *Accountability*—Having the ability to find/trace the responsible parties for actions performed
- *Privacy*—Following applicable privacy laws[1] and keeping collected personal information confidential and safe from misuse
- *Digital signature*—Signing digital documents, where both the document and the signature are integral and verifiable

[1] Regulations protecting a person's right to be left alone, and governing collection, storage, usage, and release of the individual's financial, medical, and other personal information.

Take person A sending a message to person B as an example. Authentication relates to verifying the identities of A and B. Integrity relates to the message not being changed in any way when received. Availability means the message won't be lost. Confidentiality means the message has not been seen by anyone other than A or B. If the message is signed by A, then it is not possible for A to later deny sending the message.

These concepts are also goals to achieve for cybersecurity. We'll see how various security measures can be applied to achieve these goals.

A practical understanding of security measures and technologies will be helpful for anyone in the increasingly digital world.

7.1 Login

Login is the most common form of user authentication. You log in to your computers, tablets, and smartphones so only you can use them. Personal devices may require only a password or a PIN to log in. Sometimes fingerprint or face recognition provides an alternative way for login. After login, it is important to not leave your personal computing device unattended. Lock the screen if you need to walk away briefly. Make sure to log out when you are done.

Often the first thing you do on your computer is to check email. Again, you need to log in to your email account before you can send and receive email. It is exceedingly important that no one else is able to receive/read your email. So, guard your email password carefully.

7.1.1 Website Login

Login is also required for shopping and other business on the Web. Websites often have areas and services reserved for members who are registered or have accounts with the site/organization. Each user often must log in to gain access to member-only or account-specific information or services. Access to a login page and all pages under login control normally uses HTTPS (Secure HTTP; Section 7.2) to protect the user ID and password from eavesdropping en route in the network. Section 7.2 provides more details on HTTPS.

A website requiring user login normally also provides all of the login-related functionalities, as indicated by Figure 7.1. Therefore, user authentication is only one piece of the login puzzle.

A system checks the user ID and password against securely stored values, usually in encrypted form, to perform authentication. Repeated failure to log in can result in a user being locked out of the system.

If the the user ID and password entered check out, then the user has been authenticated and will be authorized to access protected resources, collectively known as a *security realm.*

Authentication authorizes entry into a security realm consisting of a collection of protected resources. Unlike a locked room or safe, online security

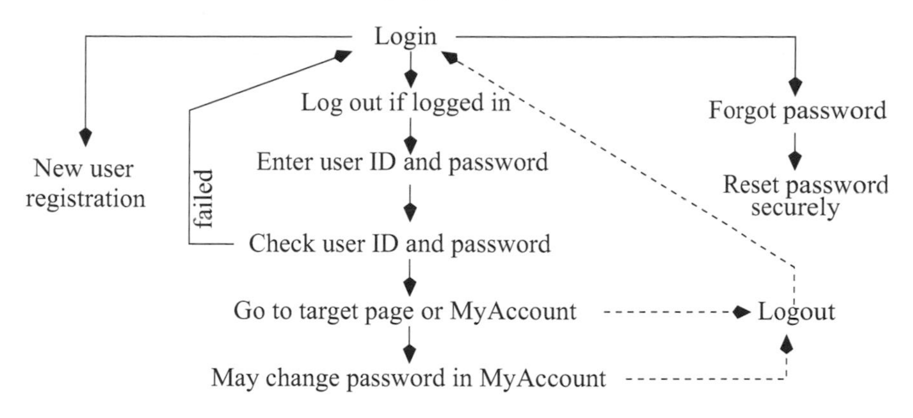

FIGURE 7.1 Login Logic Flow

realms are not something you can see or touch. But, in the digital age, we must consider and treat them the same way.

> **CT: Safeguard Security Realms**
>
> *Guard security realms the same way you would a physical safe.*

Take security realms as seriously as you would any real safety box. The consequences of an online break-in is just as damaging, if not more, as a physical one.

On a shopping site such as `Amazon.com`, for example, a customer's security realm includes his/her personal profile, account status, orders, order history, saved credit card information, billing and shipping addresses, and so on. On the Web, users sometimes will access a protected resource before login. The site will ask you to login and redirect you to the target resource automatically upon authentication. For example, clicking on "Your Amazon.com" would bring up a login page if the user has not logged-in yet.

While elaborate websites such as `amazon.com` uses login pages and programs of its own design, other sites may use the much simpler *HTTP Basic Authentication* supported by Web servers and browsers. With HTTP Basic Authentication, a generic browser-supplied login dialog box pops up for login. For example, the CT site offers interactive demos for owners of the textbook and each owner must register to obtain a personal user ID and password to access these demos. Figure 7.2 shows a browser login box for the CT site.

Pay attention when filling such generic pop-up login dialog boxes. Note the displayed website URL and realm name, `CT Demos`, in this case. The same

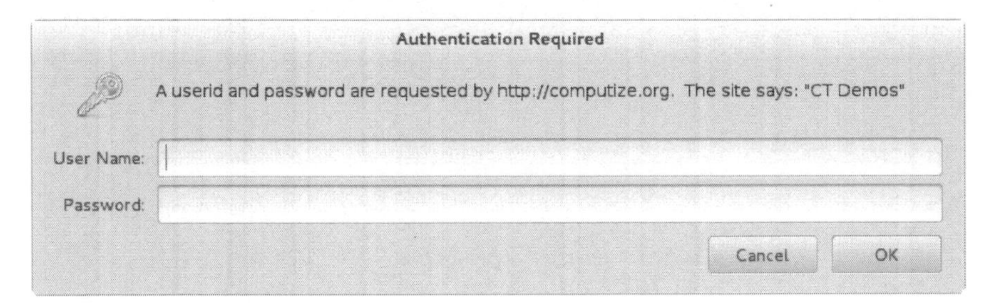

FIGURE 7.2 CT Website Login

website may have different realms. Same as using the wrong key for a door, entering login information for one realm to enter another realm will surely fail.

Login protection is as good as the quality of your password. Here are some ideas related to passwords:

- Avoid using the same password for different systems or realms. This way, if a password is compromised, you have limited the damage to one system or realm, hopefully.
- Choose passwords 8 or more characters long. Use uppercase and lowercase letters, numbers, punctuation marks, and other symbols. Keep them easy to remember but hard to guess. Never use 1234, 0000, or whole words.
- Write down your user ID , password, and other security information (such as security images and questions) somewhere safe. Consider saving them in an encrypted file (Section 7.4.1).
- Make sure you are not being observed or video recorded when you login. This is especially important when you are in a public place. Consider login to important places only from the privacy of your home.
- Change your passwords from time to time just to be extra safe.
- Do not leave your device unattended after login. Always log out immediately after finishing your business. Close your browser or shut down your system afterwards.
- Use the browser auto-login feature, where your browser remembers your user IDs and passwords for different websites, only if you enable a *browser master password* to protect the saved login information from others who may gain access to your computer. Select your browser's advanced security option to set your master password.

When conducting important business on the Web, such as banking, security

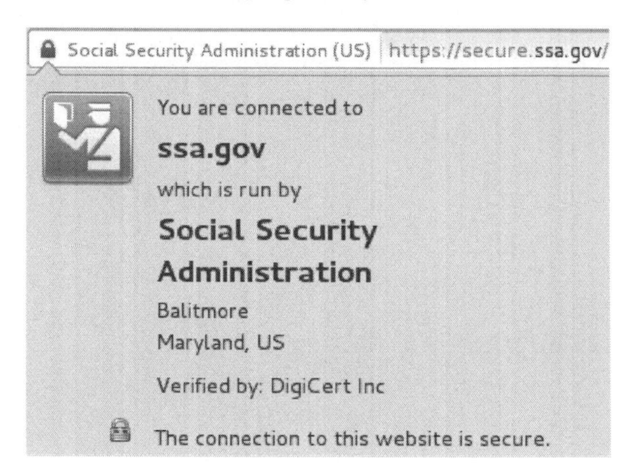

FIGURE 7.3 Sample Web Certificate Display

trading, or other official business, you need to be sure that the site is authentic and uses HTTPS for all your login and subsequent transactions. Pay attention to the browser location box. You need to see `https` in the URL and a padlock icon in front (CT: PAY ATTENTION TO DETAILS, Section 4.6.1). Click the icon to examine the *digital certificate* (Section 7.3) of the website. Figure 7.3 shows the Firefox display of URL and digital certificate of `ssa.gov`, run by the US Social Security Administration.

> **CT:** PREVENT ILLICIT LOGIN
>
> *Safeguard all your user IDs and passwords. Be vigilant with your login session.*

Computers and server hosts will, of course, require login. But it is up to each user to take login seriously. First of all, always use strong passwords and different ones for different realms. Keep user IDs and passwords confidential. Make sure no one is watching as you login. Avoid public places where there may be video surveillance or hidden cameras. Insist on HTTPS and check the server certificate. Log out (from the site) and close the browser when you are done. Shut down (log out from) the computer if it is a public one.

After login, do not leave your computer or your login session unattended. Want to get a cup of tea or go wash your hands? Lock your screen before leaving. Always carry your laptop, tablet, and smartphone with you. Don't leave any behind or in your car. When leaving work, going out, or retiring for the day, shut down your computer, turn off your monitor, or at least log off. We must be vigilant if we are to maintain security.

7.2 HTTPS and SSL/TLS

Web servers support HTTPS for secure communication between the client and the server.

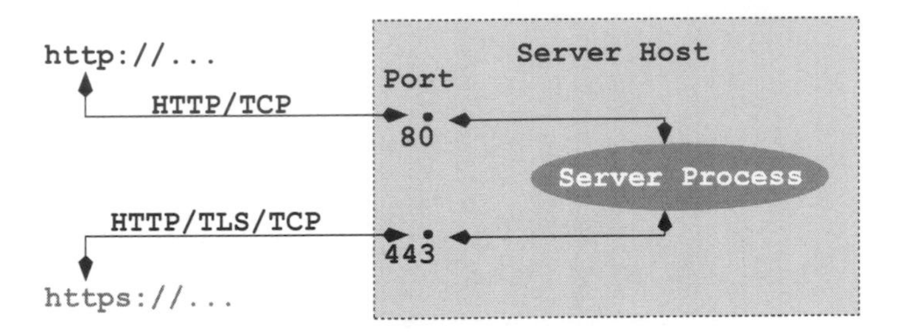

FIGURE 7.4 HTTP and HTTPS

HTTPS is HTTP (Hypertext Transfer Protocol) over *Secure Socket Layer* (SSL) or the newer *Transport Layer Security* (TLS) protocol (Figure 7.4). Note HTTP and HTTPS use different server network ports, normally 80 and 443, respectively. SSL/TLS developed from SSL 1.0, 2.0, and 3.0 to TLS 1.0, 1.1, and 1.2. SSL/TLS provides secure communication between client and server by allowing mutual authentication, the use of digital signatures for integrity, and data encryption for confidentiality. To enable HTTPS, a server needs to install a valid Web server certificate (Section 7.3) and enable SSL/TLS.

SSL/TLS may be placed between a reliable connection-oriented transport protocol layer, such as TCP/IP, and an application protocol layer, such as HTTP (Figure 7.5).

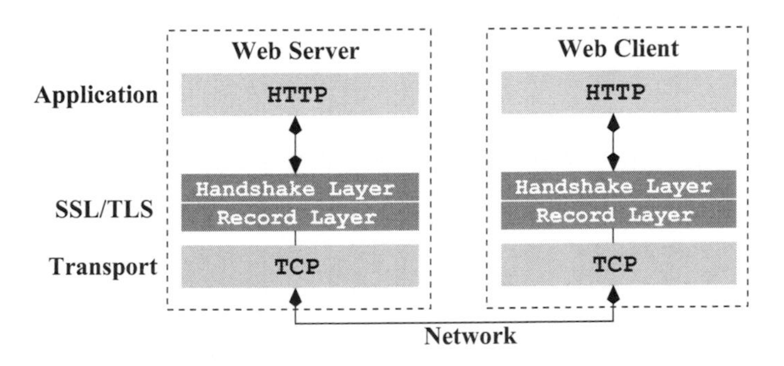

FIGURE 7.5 HTTPS Protocol Layers

Basically, TLS sets up secure communication in two steps:

1. The *handshake phase*—Mutual authentication and securely agreeing upon a randomly generated *session key* to be used in the next phase

2. The *session data phase*—Following the Record layer protocol, using the

session key for symmetric encryption (Section 7.4.1) of messages between the client and server

The handshake phase uses *public-key cryptography* (Section 7.5) for security, while the session data phase uses the more efficient symmetric encryption for speed. Each new SSL/TLS connection will establish a new session key. Figure 7.6 illustrates the TLS handshake process from a user viewpoint.

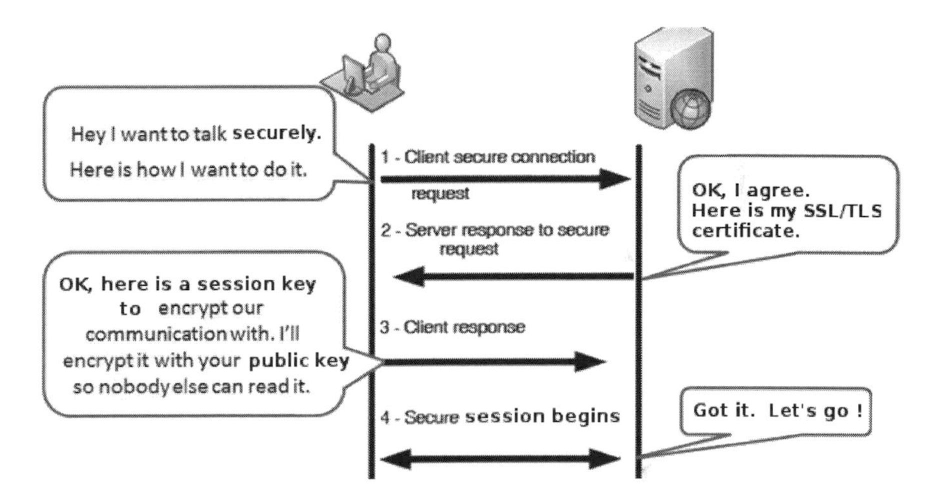

FIGURE 7.6 Basic TLS Handshake

All this is a bit overwhelming for beginners. Don't worry, we will talk about cryptography, digital signature, and all that a bit later in this chapter. But first, let's look at the digital certificate.

7.3 What Is a Digital Certificate?

In secure communication, the very first concern is that the parties are actually who they say they are. A *digital certificate* is a document (computer file) signed by a *certificate authority* (CA) that can vouch for the identity of the certificate holder. A CA is usually a well-established third party that is in the business of verifying credentials and issuing certificates digitally signed by the CA. Certificates can be issued for different purposes and different domains: Web server, email, digital signature, payment systems, and so on. SSL/TLS certificates are widely used by Web servers to enable HTTPS access. The largest CAs include Symantec[2], Comodo Group, Go Daddy, Thawte, and GlobalSign.

A CA issues a digital certificate for a customer after carefully verifying

[2]Symantec acquired VeriSign's Security Business.

the identity and legitimacy of the person or organization (the client). Each certificate is a digital ID and contains the identity of the client, the client's public key (Section 7.5), the expiration date of the certificate, and details of the issuing CA. A digital certificate is often issued for a certain specific purpose and is installed in applications that use it for that particular security purpose.

This certificate has been verified for the following uses:

SSL Client Certificate

SSL Server Certificate

Issued To

Common Name (CN)	secure.ssa.gov
Organization (O)	Social Security Administration
Organizational Unit (OU)	<Not Part Of Certificate>
Serial Number	0D:91:C7:6E:43:6E:57:19:80:F7:BB:A3:98:DB:F6:E6

Issued By

Common Name (CN)	DigiCert SHA2 Extended Validation Server CA
Organization (O)	DigiCert Inc
Organizational Unit (OU)	www.digicert.com

Period of Validity

Begins On	7/14/2014
Expires On	9/30/2016

Fingerprints

SHA-256 Fingerprint	BF:98:CC:24:AF:39:0E:A0:75:24:32:1C:0D:07:AB:CE: E8:5E:34:39:87:0E:CA:5F:BD:44:94:37:4B:A7:D5:C4
SHA1 Fingerprint	13:98:7E:5C:1A:13:62:ED:7D:CD:00:DB:AC:60:06:10:96:F4:E2:95

FIGURE 7.7 Sample Web Server Certificate Details

Figure 7.7 shows some details of the Web server certificate used by the US Social Security Administration on its `secure.ssa.gov` site. Digital certificates follow standardized formats, for example, X509v3, and are used by security programs within applications, such as Web servers, Web browsers, and email clients. Servers that support HTTPS need to install valid SSL/TLS certificates. A certificate not issued by a widely recognized CA or has expired can cause a browser warning about the certificate's status, allowing the end user to accept or reject the certificate.

Certificates for CAs are issued by other CAs. A *root CA* is one that signs its own certificate. Organizations often set up their internal certification system with a root CA controlled by their own company. This way, large companies and organization can issue digital certificates for internal use, without paying fees to commercial CAs. On Firefox, for example, the `Preferences/Advanced/Certificate` menu shows a list of CAs it recognizes.

Some Web servers are configured to take personal certificates, installed in Web browsers for client authentication instead of user ID and password for login.

CA-issued personal certificates are usually used to provide secure email with encryption. Free personal email certificates are available from `comodo.com`, for example. Email clients, such as Microsoft Outlook, Windows Mail, and Thunderbird, can use personal email certificates to encrypt emails before sending to keep them secure and private. Alternatively, you may choose the open source PGP and Gnu GPG systems that use self-generated public keys (Section 7.9) for secure email.

7.4 Cryptography

Cryptosystems keep communication safe by encryption, a technique invented long before the Internet or digital computers. The concept is simple, the *plaintext*, the original message, is encrypted into *ciphertext* before communication. Only the receiver knows how to decrypt the ciphertext back into plaintext.

For example, *rot13* (Figure 7.8) is a simple letter substitution cipher where each letter in the plaintext is replaced by a letter 13 places after it, assuming there are only 26 letters and the last letter is followed by the first letter in a cycle. A rot13 ciphertext can be decrypted by applying the encryption on it again.

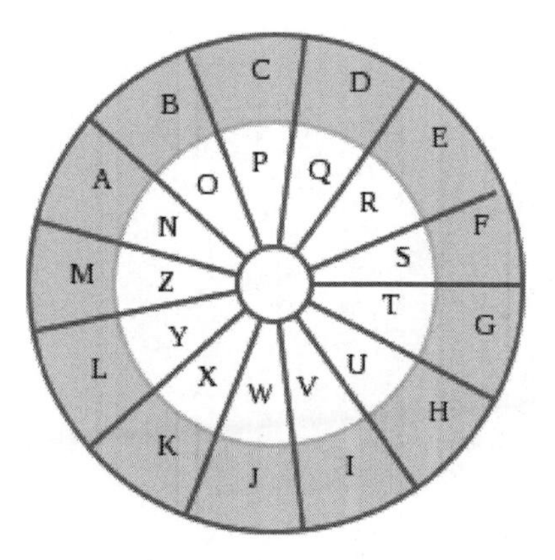

FIGURE 7.8 A Rot13 Wheel

As another example, a stationery-paper-size template with holes cut in it can be used to send secret correspondences. A sender would write the plaintext onto common stationary paper through the holes of the template. The sender would then compose an innocent sounding letter with words of the plaintext embedded in the many other words of the letter. The receiver would use a copy of the same template to easily recover the plaintext.

Or, senders and receivers can agree on a book to use. Ciphertext would contain page number, line number and word number to identify any particular word from the book. Only people who know which book and what the numbers mean can decrypt the message. Further, one of the many numbers may indicate a particular book among several possible ones to use.

Before and during World War II, the Germans made heavy use of various electromechanical rotor cipher machines known as Enigma (Figure 7.9).

FIGURE 7.9 Enigma Machine at NSA

Operators use the same machine settings to send/receive coded messages. Breaking of the Enigma code started in Poland with the cryptanalysis breakthrough made by mathematician Marian Rejewski, at Poland's Cipher Bureau, in December 1932. The work continued in France and Britain. The British Bletchley Park cryptographers, including Alan Turing, greatly advanced Enigma decryption technology with invention of the *Bombe Machine* (Figure 7.10, source `culture24.org.uk`) and helped win the war.

FIGURE 7.10 Bombe Machine in England

These are examples of *symmetric cryptosystems* (Section 7.4.1) that use the same *key* to encipher and decipher messages. Communicating parties must know the key beforehand. And the key must be kept secret to others. Obviously, rot13, the paper template, the book plus numbering scheme, the Enigma machine settings are the keys in the above examples.

Public-key cryptosystems, however, are asymmetric and use not one but a pair of keys—one to encrypt and the other to decrypt. The decryption key is kept secret, while the encryption key can be shared openly (Section 7.5).

7.4.1 Symmetric Cryptosystems

Modern electronic symmetric encryption systems (Figure 7.11) need to work on digital data. Most, if not all, of them use an encryption/decryption algorithm that is open and a key that is kept secret.

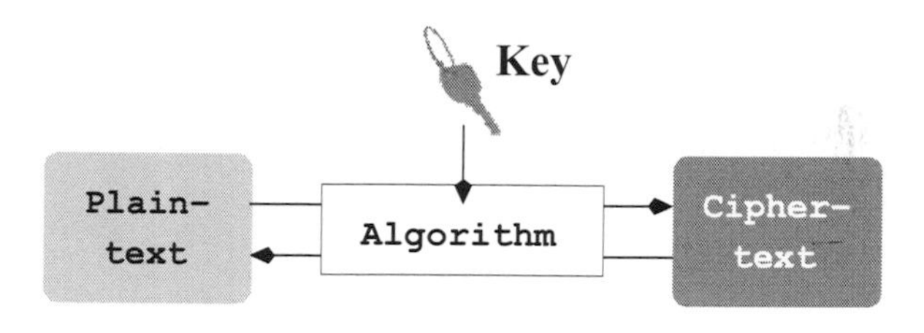

FIGURE 7.11 Symmetric Cryptosystems

- Encryption/decryption algorithm: The algorithm performs various substitutions and permutations on chunks, typically 128- or 256-bit blocks, of the plaintext or ciphertext.
- Secret key: The plaintext (ciphertext) and the secret key are input to the encryption (decryption) algorithm. The exact transformations performed depend on the key used. The algorithm produces different output depending on the key given. Using the same key on the ciphertext, the decryption algorithm produces the original plaintext.

A symmetric cryptosystem usually has these two characteristics:

1. Open algorithm: The encryption/decryption algorithm can be described in the open. It is impractical to decode any ciphertext knowing the algorithm and not the secret key.
2. Secret key: Senders and receivers must have obtained the key securely in advance and must all keep the key secret.

The secret key is usually a bit pattern of sufficient length. The quality of the secret key is important. It should be randomly generated and 256-bit or longer to make brute-force attacks, trying all possible keys, impractical.

When a password (or passphrase) is used as a key, it is usually put through a key derivation function, which compresses or expands it to the key length desired. Often, a randomly generated piece of data, called a *salt*, is also added to the password or passphrase before transforming it to the actual key.

The Advanced Encryption Standard (AES) is a symmetric cryptosystem for electronic data established by the US National Institute of Standards and Technology (NIST) in 2001. AES has been adopted by the US government and is now used worldwide. It supersedes the Data Encryption Standard (DES), which was published in 1977. There are other symmetric ciphers, such as RC4 and Blowfish, but AES-256 seems to be the best.

Let's take a closer look at AES-256, which uses a 256-bit key and encodes data by encrypting one 256-bit block at a time. The following is an over-simplified view of how it works:

1. Arranges the data block to be encoded/decoded into a 4 by 4 array of bytes
2. Generates *round keys* using the given key
3. Transforms each byte by bitwise **xor** with a *round key*
4. Scrambles and transforms the 4 by 4 array, in multiple rounds, by shifting rows, mixing columns, and substituting bytes from a look-up table derived from the current round key

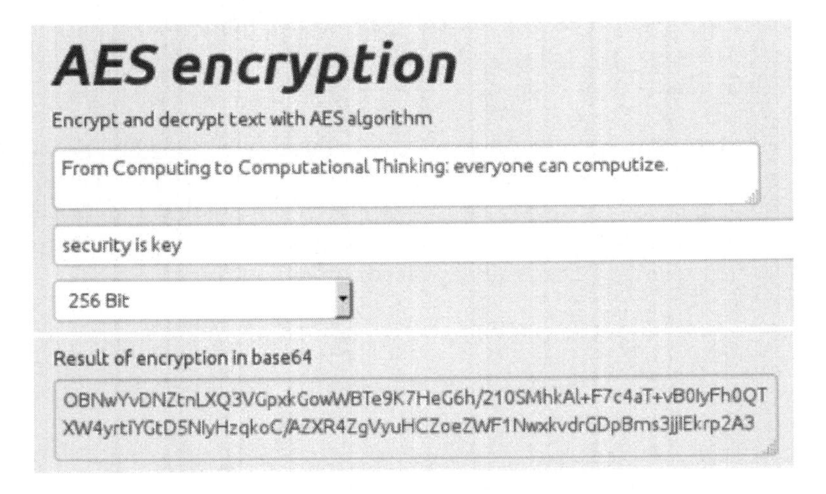

FIGURE 7.12 Sample AES Encryption

Figure 7.12 shows a plaintext involving the title of this book and the AES-256 produced ciphertext using the key "security is key." The encrypted binary result, a sequence of bytes, is displayed as a string of characters using *base64 encoding*. Base64 encoding is widely used to encode email attachments. Sixty four ASCII characters are used to represent each 6-bit piece of the binary data to make it textual, for display, printing, or email. You can try AES encryption/decryption at the CT site (**Demo:** `TryAES`).

CT: Secure Sensitive Files

Use a secret key to encode your sensitive files to keep them safe.

Files of bank statements, login information, business contacts, tax returns, accounting, financial, and insurance records, contracts, medical history, and so on should be kept confidential by encrypting them. Furthermore, it is advisable to keep scanned images of your important documents, such as birth certificates, passports, and driver licenses, in files for easy usage. But, make sure these are protected, too, by encryption.

With Microsoft Word™ or Microsoft Office™, you can save and retrieve encrypted files. The free Vim editor (`vim.org`) can encrypt/decrypt a file using Blowfish and a secret key. The AES Crypt application (`aescrypt.com`) simply encrypts/decrypts a file using AES and your own key. Both tools work on multiple platforms. Similar file encryption apps are available on smartphones.

Recording all your passwords and secret keys in an encrypted *keyfile* means you can access login information and secret keys if you just remember one key, the one for the keyfile. This can make life much easier.

CT: Add Security Layers

Establish multiple layers of defenses for stronger security.

One cannot access a computer or a security realm without login. Even after login, important tasks, such as installing new applications or creating new user accounts, require administrative privileges. Files have access permission settings. Sensitive files can be encrypted. Master passwords can be required to launch Web browsers and email clients that remember passwords for you. Firewalls and antivirus programs guard again security risks. On routers, turn on firewall, WPA2/WEP wireless security, and turn off remote router admin. You may even consider encrypting your hard disks. Multiple layers of protection can make it harder to break your security.

As mentioned in Section 7.2, symmetric cryptography is also used to keep TLS communication confidential. But, TLS still needs a secure way for key exchange and mutual authentication. That brings us to *public-key cryptography*.

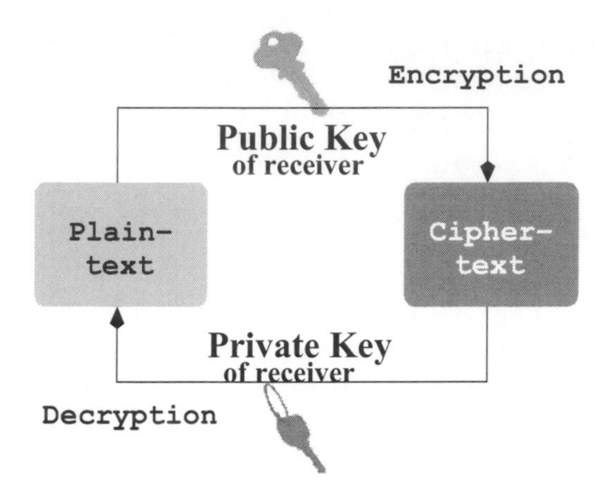

FIGURE 7.13 Public-Key Cryptography

7.5 Public-Key Cryptography

In a critical departure from symmetric cryptography which uses the same key for encryption and decryption, *public-key* cryptography uses a pair of mathematically related keys—one key, the public key, to encrypt and the other, the private key, to decrypt. The pair of keys are integers satisfying well-defined mathematical properties and usually produced by a key generation program. For each public key, there is only one corresponding private key, and vice versa.

The public key is made available for anyone who wishes to send an encrypted message to a recipient who uses the private key to decrypt the message.

The public key usually becomes part of a digital certificate, which verifiably associates the key to its owner. Or, the owner of the key pair may publish the public key in online *key repositories* open to the public. Thus, anyone who wishes to send secure messages will use the public key for encryption. The owner then uses the private key to decrypt (Figure 7.13).

> **CT:** Breakthrough
>
> *Break out of the mold of conventional thinking. Challenge old assumptions. Think outside of the box.*

Public-key cryptography, as it stands now, started with the breakthrough idea, credited to Bailey W. Diffie, Martin E. Hellman, and Ralph C. Merkle, that two parties can establish a secret symmetric key over an unprotected public communications channel. It is infeasible for any eavesdropper of their open exchanges to deduce the key they agreed upon.

When Merkle first proposed research into the idea, it was met with skepticism and resistance from experts. But, this mid-1970 invention soon led to the even more astounding public-key cryptography, where the encryption algorithm and key can be published openly.

One critical element of public-key cryptography is the computational infeasibility to deduce the private key from the public key. Unlike symmetric encryption, public-key encryption has no *secure key distribution problem.* Thus, anyone and any system can send a secure message to a receiver, who has a published public key or a digital certificate, without having any prior contact with the person.

However, the speed of public cryptography is much slower than symmetric cryptography. This explains why SSL/TLS uses public-key encryption only in the handshake phase to establish a symmetric *session key* for the actual data transmissions.

Public-key cryptography can provide not only secure symmetric key distribution, but also digital signature. The technology underpins modern cryptosystems, applications and protocols, such as digital certificates, TLS (Section 7.2), and PGP (Section 7.9).

Public-key cryptosystems were introduced by Whitfield Diffie and Martin Hellman in 1976. Commonly used public-key cryptosystems include RSA and ECC (Elliptic Curve Cryptography). RSA was developed by Ron Rivest, Adi Shamir, and Leonard Adleman in 1977. RSA (patented by RSA Data Security, Inc.) is based on the intractable problem of factoring large integers that are products of two large primes.

ECC is a more efficient public-key cryptosystem in that smaller size keys can be used. For example, a 256-bit ECC key is equivalent to a 3072-bit RSA key in terms of security strength. The key size directly affects the encryption/decryption speed. The idea of discrete logarithms in finite fields and their cryptographic significance was raised by A. Odlyzko in 1984. Cryptosystems based on elliptic curve discrete logarithm problem (ECDLP) were first proposed in 1985 independently by Neal Koblitz from the University of Washington and Victor Miller at IBM, Yorktown Heights, NY. Both RSA and ECC are available as part of the TLS protocol for secure handshake to establish session encryption.

With public-key cryptography, we need an arrangement to bind public keys to their owners and to publish, distribute, use, verify, store, manage, and revoke such bindings. This arrangement is known as a *public-key infrastructure* (PKI). A widely used PKI is based on digital certificates (Section 7.3) that contain public-key and owner identity information.

CT: Beware of Bugs

Bugs in computer programs are a fact of life. While better CT, on the part of programmers, can prevent many bugs, we still cannot let our guard down.

An algorithm or protocol will work, as well as its actual implementation. In 2014, the infamous *heartbleed bug* in some implementations of OpenSSL was exposed. A careless implementation of the *heartbeat* extension of TLS was the cause. While a TLS session is ongoing, a client may send a *heartbeat request* to the server to keep the TLS session alive. The heartbeat request contains some data bytes and a byte count. The server responds to a heartbeat request by sending the bytes back to the client. The heartbeat is a good feature that helps avoid closing down and then reestablishing a new TLS connection.

However, the erroneous OpenSSL heartbeat implementation forgot to check the incoming byte count against the actual number of bytes received. An attacker can send a heartbeat request with a small amount of data but a much larger incorrect byte count. The request causes the server, with the heartbleed bug, to send that many bytes from its memory, which usually contains sensitive information such as user IDs, passwords, and even the server private key.

7.6 RSA Public-Key Algorithm

Wondering about how exactly a public-key cipher actually works? Let's take a closer look at the RSA public-key cryptosystem, and how keys are generated and used in encryption and decryption.

RSA Keys

The RSA key generation algorithm:

1. Find P and Q, two large prime numbers. Let $M = P \times Q$ and $T = (P-1)(Q-1)$.

2. Choose E, such that (i) $1 < E < M$, and (ii) E and T are relatively prime (no factors in common).

3. Find D, such that $D \times E = 1 \; mod \; T$. We can find D as a by-product of computing $gcd(E, T)$, a very efficient operation.

4. Publish the modulus-exponent pair (M, E) as the public key.

5. Keep D safe as your private key.

As an example, let's work with very small numbers to examine the RSA key generation process:

1. Let $P = 7$ and $Q = 11$. Then we have $M = 77$ and $T = 60$.

2. We will use $E = 37$, which is relatively prime to $T = 60$.

3. We find $D = 13$, satisfying $D \times E = 13 \times 37 = 1 \ mod \ 60$.

4. The RSA key pair is [(77,37); (77,13)].

Thus, you publish the public key (77, 37) and keep (77, 13) private, or the other way around. See **Demo:** `ToyKey` and **Demo:** `KeyGeneration` at the CT site for a hands-on experiment.

In practice, much larger P and Q are needed to give a good size M for strong security. Usually M needs to be 2048 or 4096 bits in length.

RSA Enciphering and Deciphering

Modular arithmetic is used in encryption and decryption, with the modulus set to $M = P \times Q$. Assume that the public exponent is E and the private exponent is D.

1. Take a small enough block of bits from the plaintext and treat it as an integer Pt, satisfying $Pt < M$.

2. Encrypt with $Ct = (Pt^E) \ mod \ M$. Ct is the ciphertext.

3. Decrypt with $Pt = (Ct^D) \ mod \ M$.

This is because it can be proven mathematically that $Ct^D = Pt^{(E \times D)} = Pt \ mod \ M$, always. Thus, the same powering $modM$ procedure is used for encryption and decryption.

Now, we can try RSA encryption of the sample plaintext $Pt = 42$ (US ASCII for `B`) using the toy key [$M = 77$, $E = 37$]. The ciphertext is computed as follows:

$$Ct = 42^{37} = 70 \ mod \ 77$$

To decrypt the ciphertext Ct, we perform the following:

$$Pt = 70^{13} = 42 \ mod \ 77$$

Perform the above actions with **Demo:** `TryRSA` at the CT site.

The RSA public and private keys play symmetric roles. This means, if we first encrypt a plaintext using the private key, the resulting ciphertext can be decrypted using the public key. Try your own keys with **Demo:** `RSAAction` at the CT site. This symmetry makes digital signature work, as we will see next.

7.7 Digital Signature

CT: Sign Digitally

You can attach your own digital signature to a document and send the signed document to others in a secure and verifiable way.

A public-key cryptosystem, such as RSA or ECC, lends itself immediately for digital signature (Figure 7.14). A sender signs a document by encrypting it with his/her private key. The encrypted document can be sent to anyone. The receiver can decode the document using the sender's published public key. Because only the sender has knowledge and use of his/her private key, the document is verifiably produced by the sender, even in a court of law. Furthermore, the integrity of the original plaintext is also maintained.

Usually, situations requiring digital signature do not involve secrecy. Software distribution is a good example. Generally, there is no need to apply signature to the entire message or file, which can be large. Instead, signing a digest of the message or file is enough, as we'll see next.

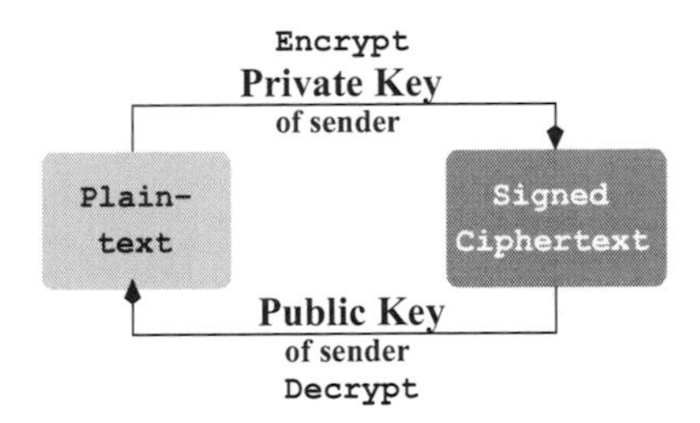

FIGURE 7.14 Digital Signature

7.8 Message Digests

Various algorithms have been devised to take a message (file) of any length and reduce it to a short, fixed-length *hash* or *digest*. Hash algorithms are designed to produce a different digest if any part of the message is altered. A *message digest* serves as a *digital fingerprint* for the entire original data.

Using well-designed algorithms, a hash function scrambles, mixes, mashes, and reduces a given message into a fixed-lengh result. The hash function is *oneway*, because it is almost impossible to deduce the original message from the digest or to find a message having the same digest. However, because

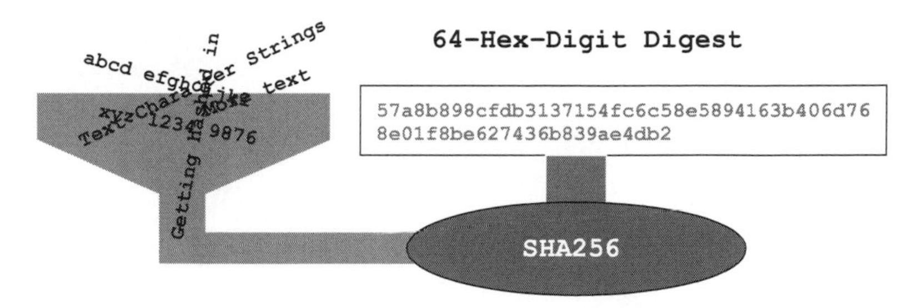

FIGURE 7.15 SHA256 Message Digest

there are an infinite number of possible messages but only a finite number of different digests, vastly different messages may produce the same digest.

The SHA suite of hash functions are part of the US Federal Information Processing Standard (FIPS). A message digest is usually displayed as a sequence of hexadecimal digits (Figure 7.15). Other digest algorithms include the older MD5, which is not as strong as SHA.

Message digests are therefore useful in verifying the *integrity* of files. Integrity is preserved if the file stays the same, unchanged, and unaltered. When software is distributed online, a good practice is to include a fingerprint for the software and a code signing certificate. Such a package allows the software download application to identify the software publisher and check the integrity of the download (Figure 7.16).

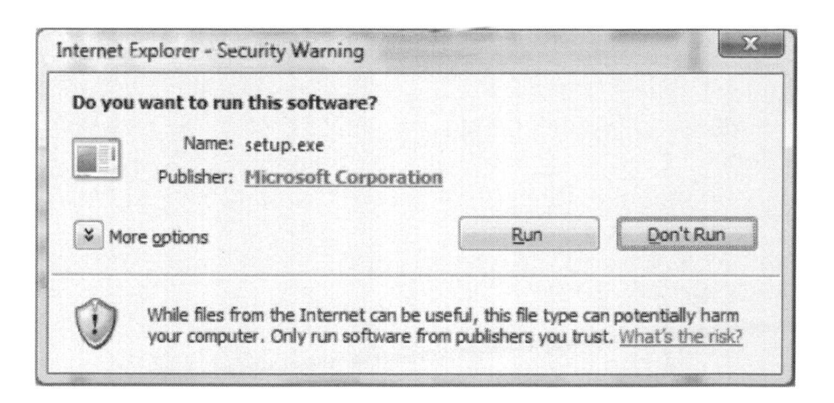

FIGURE 7.16 Secure Software Download

Attaching Signatures to Messages

To digitally sign a particular message (or file) without having to encrypt that entire message is often desirable. To do this, a digest of the message or file

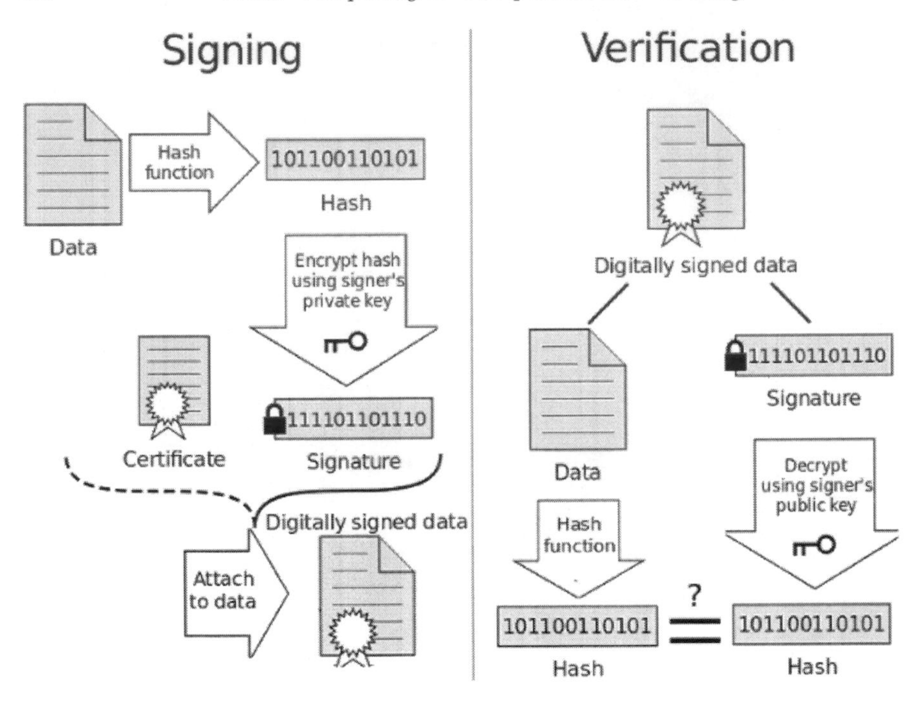

FIGURE 7.17 Attached Digital Signature

is created first, using a suitable message digest hash function (Figure 7.17; source: Wikipedia).

The digest is then encrypted using your private key to produce a signed version, which is attached to the plaintext message, together with the signer's digital certificate containing the signer's public key and the hash function. A receiver can

1. decrypt the signed message digest;
2. produce a digest of the received plaintext message using the indicated hash function; and
3. compare the two results for a match.

A match verifies the integrity and the authenticity of the received message or file.

Quantum Key Distribution

In more recent developments, quantum mechanics is used for key distribution (QKD) between two parties. QKD guarantees secure communication by ensuring a random secret key, which can then be used to encrypt and decrypt messages. Because QKD enables the two communicating users to detect, through quantum effects, any eavesdropping, it can guarantee communication security.

7.9 Secure Email

Modern email clients, such as Microsoft Outlook and the open source Thunderbird, support secure email specified by the S/MIME (Secure Multipurpose Internet Mail Extensions) standard. Once set up, you can send and receive encrypted email, as well as signed messages. When an email message is encrypted, email contents and attachments are turned into ciphertext. Of course, the email subject or other email headers are not encrypted.

When an email is signed, nothing is encrypted, except a signed message digest is attached. Normally, we do not want to sign our email. But, it is possible to both encrypt and sign an email.

The prerequisite for S/MIME is that each correspondent must have an email certificate installed in a secure email client.

Commercial personal email (S/MIME) certificates are easily available from CAs. You may even find CAs that offer free email certificates. But the certification application and verification process may be complicated and bothersome to many users.

A good alternative to S/MIME is PGP/MIME (Pretty Good Privacy) which does not require a CA-issued certificate. The free *GnuPG* (GNU Privacy Guard; GPG) is an implementation of the OpenPGP standard. Using GPG, you can generate and distribute you own key pair used for secure email.

GPG is widely available and runs on all major platforms including Mac OS X, Linux, and Windows (Gpg4win). GPG is a command-line tool but GUI (graphical user interface) versions (called GPA) are available. Download and installation are free.

Let's take a look at how to set up secure email with GPG on Thunderbird. Other email clients can be set up using a similar approach.

7.9.1 Secure Email with Thunderbird

Let's help you set up Thunderbird with OpenPGP for secure email. First, you need to install the Thunderbird email client for your system. Just go to `mozilla.org/en-US/thunderbird/` to download and install. After that, open Thunderbird and configure it to work with your email provider, be it your ISP, Gmail, Hotmail, or Yahoo! mail. If you already use Thunderbird, go to the next step.

Follow this simple procedure to enable Thunderbird secure email (PGP/MIME).

1. Download and install GPG (`gnupg.org` or `gpg4win.org`).

2. Open Thunderbird and use the `tools->Add-ons` option to search for and install the *Enigmail* add-on.

3. Follow the Enigmail set-up wizard to set up your key pair (use your correct name and email address). The email address should correspond to your Thunderbird email identity. You may choose for the key to never expire. Also it is recommended that you choose the 4096 RSA/RSA key from the `Advanced` options.

4. Optionally, you may choose to associate a JPG image with the key. This can be done later also.

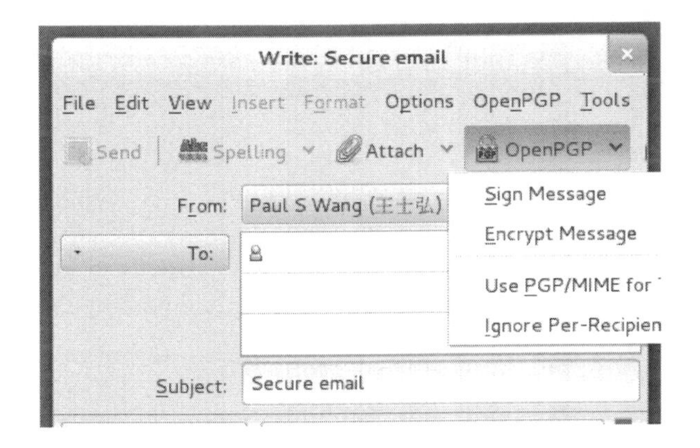

FIGURE 7.18 Thunderbird with Enigmail Added

Now, your Thunderbird is set up. But, before you can send and receive encrypted email, you need (A), to send your public key to people you know so they can send encrypted email to you; and (B) to receive/install their public keys so you can encrypt email to them.

For (A) do this:

A1 From the Thunderbird Menu (click the three-bars icon on the right-hand side of the Thunderbird menu bar), select `OpenPGP->Key Management` to pop up the Key Management dialog (Figure 7.19).

A2 In the Key Management dialog, check the option box `Display All Keys by Default`. You should see your key listed.

A3 Right click the key you want, and select the option `Send Public Key by Email`. The same option is also available from the `File` menu.

A4 An email composition window opens with the key file (with `.asc` suffix) already attached. Just send this email normally. That is it.

For (B), do this:

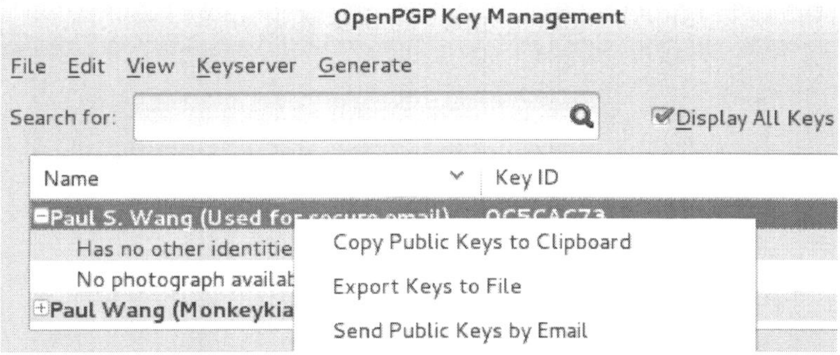

FIGURE 7.19 Sending Public Key

B1 Open incoming email from your friend containing his/her public key as attachment.

B2 Open (double click) the key file attachment (with `.asc` suffix), and Thunderbird will install the public key received automatically.

Now you are truly ready for secure email. After composing your message and adding attachments, select the options `OpenPGP->encrypt` (to encrypt the email message) and/or `OpenPGP->sign` (to add a signature) before sending. If the email has one or more attachments, be sure to also select the `OpenPGP->PGP/MIME` option. When you open an encrypted email in your inbox, Thunderbird asks you for the passphrase of your private key and then decrypts it for you automatically.

Enigmail may not be available yet on mobile devices. For Android devices, the APG (similar to GPG) app and K-9 email app combine to provide secure email following the OpenPGP standard. On iPhones, try the iPGMail app.

Also, the Signal app keeps your iPhone calls and texts private. On Android phones try the TextSecure app.

CT: Free from Surveillance

Set up secure email now. You may not need it immediately. But it should be an option for others to communicate with you.

Pulitzer Prize (2014) winning journalist Glenn Greenwald almost did not get connected with NSA (US National Security Agency) whistleblower Edward Snowden, because Greenwald had not set up PGP for his email. In "*How I Met Edward Snowden*," he recounted how Snowden wanted him to set up PGP:

> "There are people out there you would like to hear from, but they will never be able to contact you without knowing their messages cannot be read in transit."

Not everyone will be involved in such high-stakes situations. Still, it borders on inconsideration in the digital age for one not to give others the option to communicate securely.

7.10 Security Attacks and Defenses

In 2013, one day after Thanksgiving, Target, a large retail chain in the USA, had a security breach. According to a *New York Times* blog, "Cybercriminals appear to have focused on the point-of-sale systems in Target's retail stores, which collect information from customers' credit and debit cards, and potentially personal identification numbers, or PINs." The stolen information can be used to create counterfeit credit or debit cards.

In 2003, a *slammer worm* infected 90% of vulnerable computers within 10 minutes. This caused interferences with elections, airline flights cancellation, failure of Seattle's 911 emergency system, and failure of more than 13,000 Bank of America ATMs. The slammer worm denial of service attack exploited a buffer overflow bug in Microsoft's flagship SQL Server and Desktop Engine database products, for which a patch had been released six months earlier.

There were a number of other widely known attacks. Cybersecurity attacks can be from a single individual or a well-organized group, some, the so-called *advanced persistent threat* groups, could be connected to industry or even governments.

Generally, a cybersecurity attack exploits one or more vulnerabilities in your system or network, including the Internet as well as phone networks. Security vulnerabilities often are due to flaws in system design or software coding, but can sometimes be the result of software/hardware features and capabilities. And, most disturbingly, they can often be the result of carelessness, laziness, or lack of awareness on the part of system operators or end users. Let's take a look at some common kinds of attacks.

- Spoofing—Pretending to be someone, at some IP address, from a certain website, sent from some email address, or located at certain GPS locations. Spoofing is usually done by falsifying data used in communication protocols. For example, the email sender (the `From` header) can be spoofed easily.

- Phishing—Collecting private or confidential information such as user IDs, passwords, Social Security Numbers, driver's license numbers, account numbers, phone numbers, PINs, addresses, and birth dates by tricking users to supply them through phone calls, emails, or fake websites. For example, an email may ask the user to increase email storage space, change login information, or fix an old unpaid invoice by clicking a link in the email. The link leads to an official looking online form put up by the attacker. Or, a scam may inform you of a sudden wealth that you can receive by sending your bank account information and often a handling fee or tax!

- Malware—Malicious software of all kinds including computer viruses, ransomware (for example, the infamous CryptoLocker), worms (spreading themselves through the network), trojan horses (hiding in seemingly legit applications), keyloggers, spyware, and rogue security programs. The majority of active malware threats are usually worms or trojans rather than viruses.

- Backdoors—Ways to stealthily access a computer without being authenticated or defeat an otherwise secure cryptosystem. For example, a *rootkit* is malware installed deep in the operating system that enables intrusion and privileged program execution without leaving an audit trail.

- Eavesdropping—Spying by secretly monitoring network communications or leaking emissions from equipment. The man-in-the-middle attack carries this further by intercepting messages between two correspondents, and perhaps even altering the messages as they are passed along to the other end.

- Illicit direct access—Obtaining physical possession or direct manipulation of computing devices that are to be compromised.

- Denial of service (DoS)—Overloading a system or network so it won't be able to function normally and render its intended services.

Cyber crimes are a serious and global concern. Governments, private sectors, and academic institutions have acted to produce countermeasures, including legislation, regulation, law enforcement, protection of communication infrastructures, and *Computer Emergency Readiness Teams* (CERTs) in the USA and other countries.

CT: All for One and One for All

Do our best and we, the users, developers, and providers, can collectively make our shared cyberspace more secure.

These organized countermeasures and the technologies for identification, authentication, encryption, and so on are all well and good. But, the human factor is still the weakest link in cybersecurity. The 1978 robbery of the Security Pacific National Bank in Los Angeles by Stanley Mark Rifkin is a case in point. As a computer consultant for the bank, Rifkin learned the wire transfer procedures. And, on Oct. 25, 1978, he walked into the wire transfer room and memorized the secret daily fund transfer code, which was posted on the wall (surprise!). He managed to have 10.2 million dollars transferred to his Swiss bank account and purchased Russian diamonds which he immediately smuggled back to California. Of course, he got caught later, and the rest is history.

As users and operators in cyberspace, we all need to do our best to tighten security, and hopefully we can collectively make it more secure for everyone. We certainly should not post our passwords on walls.

Here are some suggested actions for individual users.

- Use strong and different passwords. Save login information in encrypted files. Change your passwords from time to time. Do not give actual answers to security questions. Invent answers others cannot deduce.
- Use common sense. Be careful opening email attachments or clicking links in an email. They may be phishing or contain malware. Report any such attempts to prominent search engines and other authorities, such as `us-cert.gov`.
- Download and install software only from known and trusted websites. Avoid FREE software that is too good to be true.
- Do not give your user ID or password in response to an email or a phone call. Make sure you have initiated a login by yourself and you have examined the HTTPS displayed server certificate.
- Back up your important files on external disks (detachable from your computer) or on the cloud in encrypted form.
- On your wireless router, use WPA2/WEP wireless security and turn off remote admin access.
- Do not access your online accounts from public places or use borrowed computers.
- Keep your laptops, tablets, and smartphones with you all the time.
- Be careful with flash drives and other similar free gift items. They may contain malware that can infect your computer.
- Lock the screen if you need to leave your workstation for a chore. Always put your smartphone back in your pocket or purse. Never leave it lying around.
- Look at the URL of secure webpages, and insist on HTTPS connection with a valid secure server certificate.

- Set up encrypted email communication for privacy.
- Encrypt sensitive files on your computers and smartphones.
- Be extra careful with Microsoft IE and Outlook; most security attacks target these applications. Consider using computers running Mac OS X, or better, Linux.
- Enable firewalls and configure them correctly on your routers and computers. Close down your Web browser after finishing with a login session.
- For mobile devices, install apps only from official app stores, and enable the screen lock (and SIM card lock) features. Install an anti-virus app.
- Report email scams, phishing, Web forgery, and other security attacks you encounter immediately. Contact authorities or the legitimate businesses to alert them. Use your Web browser's `Help->Report Web Forgery` option. Send email to

  `reportphishing@antiphishing.org`.

  Or contact the Internet Crime Complaint Center (`ic3.gov`).

Do your best, and you'll be glad you did. If everyone does his/her part, cyberspace will be that much more secure.

Additionally, software and systems developers need to put security as a primary requirement and make secure applications easy to use as a rule. A free and secure cyberspace is something we all need to guard actively.

Exercises

7.1. What is the difference between identification and authentication?

7.2. What is a security realm?

7.3. What is a digital certificate? A root CA?

7.4. Consider a 256-bit key in a symmetric cryptosystem. How many different keys are there? How difficult is a brute-force attack?

7.5. What is the difference between HTTP and HTTPS? When should HTTPS be used instead of HTTP?

7.6. Find out the details about base64 encoding. Which 64 characters are used? How and why is the plaintext padded?

7.7. What is the the difference between GPG and PGP?

7.8. Follow the instructions in Section 7.9.1 and set up secure email in Thunderbird. Test it by sending and receiving secure emails from friends you trust.

7.9. Find out about the PGP *Web of trust* and compare it with CA-signed certificates.

7.10. Explain in your own words the difference between symmetric and public-key cryptosystems.

7.11. Obtain a free email certificate and install it in your email client and perform tests.

7.12. **Computize**: discuss usage of the terms "client and server", "server", "client", "server host", "client host", "server side", "client side".

7.13. **Computize**: How to keep your password safe and readily available.

7.14. **Computize**: Is texting on your smartphone safe and secure?

7.15. **Computize**: Is video calling, such as Skype and Google Hangout, safe and secure?

7.16. **Computize**: Add to the list of actions end users can do to help cyber-security.

7.17. **Group discussion topic**: *I forgot my password.*

7.18. **Group discussion topic**: *Cybersecurity in light of the Edward Snowden revelations about the NSA surveillance.*

Chapter 8

Solve That Problem

Computers are universal machines that can be programmed to perform almost any task. They allow us to automate solutions to problems. Such solutions can be performed by computers repeatedly, reliably, precisely, and with great speed. Not all problems lend themselves to automated solutions. But CT demands that we try to find solutions that can be automated.

The stages of computer automation are *conceptualization*, *algorithm design*, *program design*, and *program implementation*. In other words, programmers must devise solution strategies, implement algorithms, and write program code to achieve the desired results.

Specifying and implementing solution algorithms requires precise and water-tight thinking. Also required is anticipation of possible input as well as execution scenarios. Few are born with such talents. But, we all can become better problem solvers by studying cases, experimentation, and building our solution repertoire.

Often, there are multiple algorithms to solve a particular problem or to achieve a given task. Some may be faster than others. Some may use fewer resources. Others may be easier to program or less prone to mistakes. Analyzing and comparing different solution algorithms is an important part of problem solving.

Usually, we are after the fastest algorithm or one that uses the least amount of memory space. Problem-solving skills from computing can be useful in other areas, as well as in our daily lives.

8.1 Solving Puzzles

A good way to begin thinking about problem solving is perhaps by looking at a few puzzles.

8.1.1 Egg Frying

The Problem: A pan can fry up to two eggs at a time (Figure 8.1). We need to fry three eggs. Each egg must be fried for one minute on each side. Design an algorithm to fry the three eggs in as little pan-frying time as you can manage.

FIGURE 8.1 Egg Frying

The slowest method would fry one egg at a time, cooking each side for one minute. It would take a total of six minutes of pan-frying time.

A better method would fry two eggs, turning them over after a minute, and cook for another minute. After that, fry the third egg. This method takes a total of four minutes.

Is this the best we can do? Well, no. We can do better:

1. Start with frying two eggs for one minute.
2. Take one of the half-done eggs out of the pan, turn the other egg over, and add the third egg in the pan, fry for one minute.
3. Take out the egg that is done, flip the other egg over, and add back the half-done egg, fry for one minute.
4. Take both eggs out and terminate.

Each step takes one minute. The whole procedure takes three minutes to get the job done.

Again, is this the best we can do? Well, yes. How do we know? Can we prove that no method can be faster?

Here is a proof. The three eggs require six egg-side-minutes of frying. The pan can supply a maximum of two egg-side-minutes per minute. Thus, at least three minutes are needed to produce six egg-side-minutes, no matter how the frying is done.

8.1.2 Liquid Measuring

The Problem: We have a 7-oz cup and a 3-oz cup (Figure 8.2). Unfortunately, neither has any volume markings on it. We have a water faucet, but no other containers. Find a way to measure exactly 2 ounces of water.

We see the allowable operations are filling water from the faucet, pouring water from one cup to the other, and emptying the cups. Here is how we can proceed:

FIGURE 8.2 Two Cups

1. Fill the 3-oz cup completely.
2. Empty the 3-oz cup into the 7-oz one.
3. Repeat steps 1 and 2 one more time.
4. Fill the 3-oz cup again completely.
5. Pour water from the 3-oz cup into the 7-oz cup until it is full.
6. Two ounces of water now remain in the 3-oz cup.

What about measuring 5 or 6 ounces of water? If we switch the 7-oz cup to an 8-oz cup, can you solve the same problem? What if we switch it to a 6-oz cup? Try **Demo:** `LiquidMeasure` at the CT site.

8.1.3 A Magic Tray

The Problem: A magic tray is a perfect square and has four corner pockets (Figure 8.3) whose openings look exactly the same. Inside each pocket is a cup hidden from view. The tray also has a green light at the center.

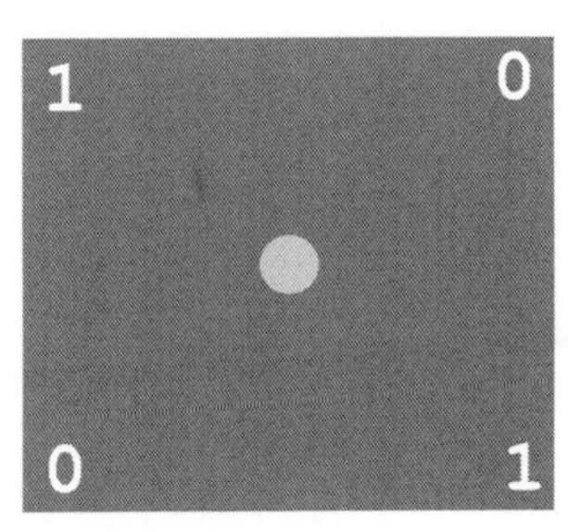

FIGURE 8.3 Magic Tray Puzzle

Cups inside the pockets can be either up (1) or down (0). The light will turn red automatically if all four cups are in the same orientation.

Your job is to turn the light red by performing a number of steps. Each step consists of reaching into one or two pockets to examine and optionally reorient the cups. No other operations are allowed. Remember, you can't see the cups. Figure 8.3 shows one possible configuration to give you an idea.

To complicate things, the tray immediately spins wildly after each step. When it stops spinning, there is no way to tell which pockets you had examined in the last step.

Your job is to create an algorithm to turn the light red, no matter what the initial orientations of the cups are. Remember, an algorithm must specify exactly what to do at each step and guarantee termination after a finite number of steps. Therefore, keep reaching into pockets and turning cups up (or down) is not a solution because you may be extremely unlucky and reach into the same pockets every time.

One observation we can make is that we may choose to reach into pockets along a side or on a diagonal. But there is a chance, however small, that we may reach into the same side/diagonal all the time. Our algorithm must work even if it never ever examines all four pockets. This puzzle is fun and hits home the algorithm ideas perfectly. We will leave the reader to work out this algorithm. It should take no more than 7 steps. See Exercise 8.1 for a hint and **Demo:** `MagicTray` at the CT website for a solution.

8.2 Sorting

In automated data processing, the need to arrange a list of items into order often arises. In computer science, to *sort* is to arrange data items into a linear sequence (in memory) so individual items can easily be retrieved. In fact, we have seen the binary search algorithm for efficient retrieval from a sorted list (Section 1.8).

Sorting is a major topic in algorithm design. Many ways have been devised and research papers published for sorting.

8.2.1 Bubble Sort

Let's present here the most basic *bubble sort* algorithm. We shall see how bubble sort rearranges the n elements of a sequence a_0, a_1, ..., a_{n-1} to put them into ascending order:

$$a_0 \leq a_1 \leq a_2 \leq ... \leq a_{n-1}$$

Bubble sort performs a number of *compare-exchange* (CE) operations. Each compare-exchange operation compares two adjacent elements of the sequence, and exchanges them if necessary so the latter element is larger. Here is the pseudo code for exchanging any two values x and y.

Algorithm `exchange(x,y)`:
Input: Integer `x`, integer `y`
Effect: Values of `x`, `y` exchanged

1. Set `temp = y`
2. Set `y = x`
3. Set `x = temp`

To describe bubble sort, let's see how it sorts a sequence of eight elements, a_0, a_2, ..., a_7. Pass 1 performs seven CE operations (Figure 8.4): CE(a_0, a_1),

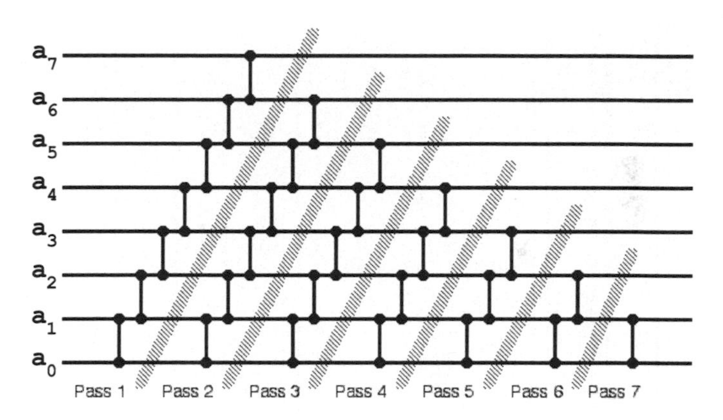

FIGURE 8.4 Bubble Sort

CE(a_1, a_2), CE(a_2, a_3), CE(a_3, a_4), CE(a_4, a_5), CE(a_5, a_6), and CE(a_6, a_7).

In a similar manner, pass 2 moves the largest value of a_0, a_1, ..., a_6 up into a_6; pass 3 moves the largest value of a_0, a_1, ..., a_5 up into a_5; and so on, until, finally, pass 7 moves the the largest value of a_0 and a_1 into a_1 to complete the sorting.

Let's put bubble sort in an algorithmic form (CT: MAKE IT AN ALGORITHM, Section 1.8). The bubble sort pseudo code, using `for` loops (Section 3.5.2), is as follows:

Algorithm `bubblesort`:
Input: Array of integers `a[0]` ... `a[n-1]` and the array length `n`
Effect: The array elements are rearranged in ascending order

```
for (int end=n-1; end > 0; end=end-1)
{   for (int i=0; i < end; i=i+1)
    {   if (a[i] > a[i+1])
        {  temp=a[i+1];  a[i+1]=a[i]; a[i]=temp;  }
    }
}
```

Note the CE operation in the body of the inner loop.

So how efficient is `bubblesort`? For the 8-element case, pass 1 performs 7 CE operations, pass 2 performs 6 CE operations, and so on. Thus, for sorting 8 elements, `bubblesort` performs:

$$7 + 6 + 5 + 4 + 3 + 2 + 1 = 4 \times 7 = 28$$

CE operations. In general, for sorting n elements, `bubblesort` performs

$$(n-1) + (n-2) + ... + 1 = \frac{n \times (n-1)}{2}$$

CE operations.

8.2.2 Improved Bubble Sort

One of the basic activities in computing is finding more efficient ways to do things and to make improvements to the procedures we already have. Take the `bubblesort` in Section 8.2.1, for example. For any given input sequence of length n, it always requires $\frac{n \times (n-1)}{2}$ CE operations, even if the given sequence is already in order!

For any algorithm, an input that causes the most number of operations is called a *worst case*. With bubble sort, the *worst case*, where the given sequence is in descending order, does require $\frac{n \times (n-1)}{2}$ CE operations. But, can we do better in less bad cases?

Sure we can. Observe that for any single pass, if no exchange is done after CE on the first two elements, then the sequence is already in order and we can terminate the procedure. Let's refer to this as the *in-sequence condition.*

We can modify the `for` loop in algorithm `bubblesort` to incorporate this improvement.

```
go_on=1;
for (int end=n-1; go_on && end > 0; end=end-1)
{  go_on=0;
   for (int i=0; i < end; i=i+1)
   {  if (a[i] > a[i+1])
      {  temp=a[i+1]; a[i+1]=a[i]; a[i]=temp; go_on=i; }
   }
}
```

A logical variable `go_on`, originally set to true, is used to control the outer `for` loop. Immediately before each pass, `go_on` is set to false. Any exchange operation where `i` is greater than `0` causes `go_on` to become true. The `go_on` being false stops the looping from proceeding to the next pass. Trace the code and satisfy yourself that it works.

Before discussing more examples, let's first give some general principles and approaches for problem solving.

> **CT: Cut It Down**
>
> *Solve a problem by reducing it to increasingly smaller and simpler problems.*

For example, bubble sort *chips away* at the problem. With each pass, the length of the sequence to be sorted is reduced by one. And repeated chipping soon gets the job done.

When solving a problem, we generally have two approaches, the *top-down* approach and the *bottom-up* approach. Top-down problem solving involves taking a big and complicated problem and breaking it down into several smaller subproblems (divide and conquer). Each subproblem can either be solved directly or be broken down further. When all the subproblems are solved, the big problem is finished.

CT: BUILD IT UP

Combine basic known quantities and methods to build ever larger and more complicated components that can be applied to solve problems.

In contrast, the bottom-up approach of problem solving starts from given data, known quantities, and methods to build up components which can be combined to achieve a particular task or problem solution. The way modern digital computers work is a good example of the bottom up approach. Bits plus their basic logic operations (logic gates) combine to represent and manipulate data, eventually leading to programs that can solve all kinds of problems. The LEGO® toy is an excellent example of the bottom-up approach.

When trying to solve a problem, we often need to apply top-down as well as bottom-up thinking. Breaking the given problem down while also combining known data and techniques up can often help us find solutions.

CT: STEPWISE REFINEMENT

Set up the overall problem solving strategy and how different parts of the solution interact/cooperate. Make it work first. Then make improvements in the various steps to fine tune the algorithm or program.

When specifying the solution algorithm or implementing a program, it is easy to get lost in the details and *not see the forest for the trees*. The *stepwise refinement* methodology in computer science suggests that we focus on the big picture and get the solution framework going correctly first. Once we have a working algorithm/program, we can find many places to make refinements for better speed and efficiency.

Adding the *in-sequence condition* improvement to the basic bubble sort algorithm is an example. We'll see more such examples later. However, we must understand at the same time, no amount of stepwise refinement can rescue an inefficient algorithm such as bubble sort.

> **CT: VERSION 2.0**
>
> *Taking advantage of user experience and feedback, software evolves and improves constantly with time. So should we in what we do in our daily lives.*

Every time we download an application update or purchase the next version of an operating system, it reminds us of the idea of "constant improvement." For companies, organizations, and individuals, the question is, "Are you improving? Where is your version 2.0?"

8.3 Recursion

A circular definition is usually no good and to be avoided, because it uses the terms being defined in the definition, directly or indirectly. For example *Bright: looks bright when viewed.* Or *Adult: person not a child* and *Child: person not an adult.* A circular definition is like a dog chasing its tail. It goes on and on to no end. However, a term or concept can be defined *recursively* when the definition contains the same term or concept without becoming circular. A *recursive definition* has *base cases* that stop the circling at the end. For people first exposed to recursion, the concept can be confusing. But, it is simple once you understand it. Please read on.

In mathematics, a *recursive function* is a function whose expression involves the same function. For example, the factorial function

$$n! = n \times (n-1) \times (n-2) \times ... \times 1, \text{ integer } n > 0$$

can be recursively defined as

For $n = 1$, $1! = 1$ (base case)
For $n > 1$, $n! = n \times (n-1)!$ (recursive definition)

In computing, a recursive function or algorithm calls itself either directly or indirectly. Here is the pseudo code for a recursive `factorial` function.

Algorithm `factorial(n)`:
Input: Positive integer `n`
Output: Returns `n!`

1. If `n` is `1`, then return 1

2. Return `n` × `factorial(n-1)`

The "return" operation terminates a function and may also produce a value. Step 2 returns the value `n` times the value of `factorial(n-1)`, which is a call to the same function itself (Figure 8.5).

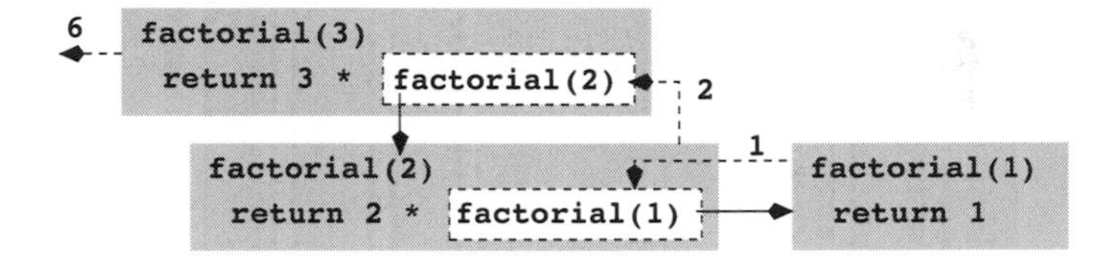

FIGURE 8.5 Recursive Calls and Returns

To solve a problem recursively, we reduce it to one or more smaller problems of the same nature. The smaller cases can then be solved by applying the same algorithm *recursively* until they become simple enough to solve directly.

> **CT: REMEMBER RECURSION**
>
> *Think of recursion when solving problems. It can be a powerful tool.*

To appreciate the power of recursion and to see how it is applied to solve nontrivial problems, we will study several examples.

Greatest Common Divisor

Consider computing the *greatest common divisor* (GCD) of two non-negative integers a and b, not both zero. Recall that $gcd(a, b)$ is the largest integer that evenly divides both a and b. Mathematics gives $gcd(a, b)$ the recursion

1. If b is 0, then the answer is a

2. Otherwise, the answer is $gcd(b,\ a \ mod \ b)$

Recall that $a \ mod \ b$ is the remainder of a divided by b. Thus, a recursive algorithm for $gcd(a, b)$ can be written directly:

Algorithm `gcd(a,b)`:
Input: Non-negative integers `a` and `b`, not both zero
Output: The GCD of `a` and `b`

1. If `b` is zero, return `a`
2. Return `gcd(b, a mod b)`

Note, the algorithm `gcd` calls itself, and the value for `b` gets smaller for each successive call to `gcd` (Table 8.1). Eventually, the argument `b` becomes zero and the recursion unwinds: The deepest recursive call returns, then the next level call returns, and so on until the first call to `gcd` returns.

TABLE 8.1 Recursion of `gcd(1265,440) = 55`

Call Level	a	b
1	1265	440
2	440	385
3	385	55
4	55	0

8.3.1 Quicksort

Another good example of recursive algorithms is the *quicksort*. Among many competing sorting algorithms, quicksort remains one of the fastest. It is much faster than bubble sort, which is not used in practice.

Let us consider sorting an array of integers into increasing order with quicksort. The idea is to split the array into two parts, smaller elements to the left and larger elements to the right. This is the *partition* operation, which first picks any element of the array as the *partition element*[1], `pe`. By exchanging elements, the array can be arranged so all elements to the right of `pe` are greater than or equal to `pe`. Also, all elements to the left of `pe` are less than or equal to `pe`. The location of `pe` is called the *partition point.*

After partitioning, we have two smaller ranges to sort, one to the left of `pe` and one to the right of `pe`. The same method is now applied to sort each of the smaller ranges. When the size of a range becomes less than 2, the recursion terminates.

Algorithm `quicksort(a, i, j)`:
Input: Array `a`, start index `i`, end index `j`
Effect: Elements `a[i]` through `a[j]` in increasing order

1. If `i` greater than or equal to `j`, then return;

[1]The `pe` is often also known as the pivot element.

2. Set `k = partition(a, i, j)`

3. `quicksort(a, i, k-1)`

4. `quicksort(a, k+1, j)`

Algorithm `quicksort` performs sorting on a range of elements in the array `a`. The range starts with index `i` and ends with index `j`, inclusive. To sort an array with 8 elements, the call is `quicksort(a, 0, 7)`.

The algorithm starts with the base case: If $i \geq j$, the range is either empty or contains a single element, and there is nothing to do and the algorithm returns. If `j` is bigger than `i`, a *partition* operation is performed to split the range into two parts, as described earlier. The `pe` has found its final position on the array at index `k`. Then, `quicksort` is called recursively to sort each of the two smaller arrays on either side of the `pe`.

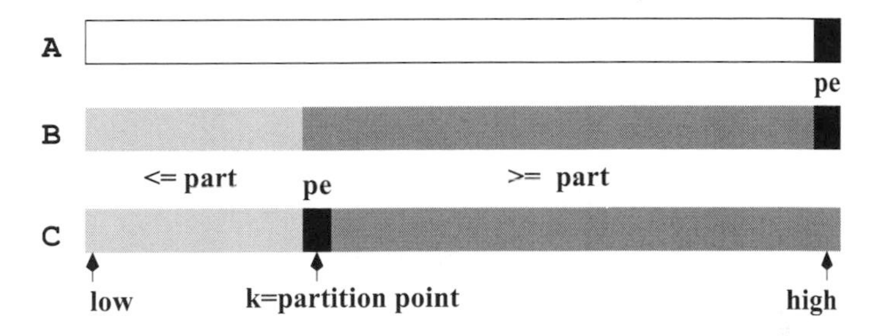

FIGURE 8.6 Partition in Quicksort

The `partition` function receives the array `a` and two indices, `low` and `high`, as input and performs these three steps.

1. Picks a `pe` in the given range of elements, randomly or by some simple criteria. Exchange `pe` with `a[high]` so that `pe` is saved at the end of the range (Figure 8.6 A).

2. Searches from both ends of the range (`low, high-1`) toward the middle for elements belonging to the other side, interchanging out-of-place entries in pairs, resulting in the configuration shown in Figure 8.6 B.

3. Exchanges `pe` with the element at the partition point (Figure 8.6 C) and returns the `pe` index (`k`).

The pseudo code, using `while` loops (Section 3.5.1), for the `partition` function, can also be given.

```
partition(a, low, high)
{  k = low; j = high;
   exchange(a, (k+j)/2, j); pe = a[high];
   while (k < j)
```

```
    { while (k < j && a[k] <= pe) k=k+1;
      while (k < j && a[j] >= pe) j=j-1;
      if (k < j) exchange(a, k, j);
    }
    if (k != high) exchange(a, k, high);
    return(k);
}
```

In the unfortunate case where pe happens to be the largest (smallest) element in the range low to high, the partition position becomes high (low), and the split is not very effective. The split becomes more effective when the two subproblems are of roughly equal size (See Exercise 8.5).

Quicksort is efficient in practice, because it often reduces the length of the sequences to be sorted quickly. Note also that quicksort performs the reordering *in place*. No auxiliary array, as required by some other sorting algorithms, is used. The best way to understand how quicksort works is to try an example with less than 10 entries by hand.

8.4 Recursive Solution Formula

CT: Apply the Recursion Magic

Answer two simple questions and you may have magically solved a complicated problem.

For many, recursion is a new way of thinking and brings a powerful tool for problem solving. To see if a recursive solution might be applicable to a given problem, you need to answer two questions:

- Do I know a way to solve the problem in case the problem is small and trivial?
- If the problem is not small or trivial, can it be broken down into smaller problems of the same nature whose solutions combine into the solution of the original problem?

If the answer is yes to both questions, then you already have a recursive solution!

A recursive algorithm is usually specified as a function that calls itself directly or indirectly. Recursive functions are concise and easy to write once you recognize their basic structure. All recursive solutions use the following sequence of steps.

(i) Termination conditions: Always begin a recursive function with tests to catch the simple or trivial cases (the base cases) at the end of the recursion. A base case (array size zero for `quicksort` and remainder zero for `gcd`) is treated directly and the function returns.

(ii) Subproblems: Break the given problem into smaller problems of the same kind. Each is solved by a recursive call to the function itself passing arguments of reduced size or complexity.

(iii) Recombination of answers (optional): Finally, take the answers from the subproblems and combine them into the solution of the original bigger problem. The function call now returns. The combination may involve adding, multiplying, or other operations on the results from the recursive calls.

For problems, like the GCD and quicksort, where no recombination is necessary, this step becomes a trivial return statement. However, in the factorial solution, we need to multiply by n the result of the recursive call $factorial(n-1)$.

The *recursion engine* described here is deceptively simple. The algorithms look small and quite innocent, but the logic can be mind-boggling. To illustrate its power, we will consider the *Tower of Hanoi* puzzle.

8.5 Tower of Hanoi

Legend has it that monks in Hanoi spend their free time moving heavy gold disks to and from three poles made of black wood.

FIGURE 8.7 Tower of Hanoi Puzzle

The disks are all different in size and are numbered from 1 to n according to their sizes. Each disk has a hole at the center to fit the poles. In the beginning, all n disks are stacked on one pole in sequence, with disk 1, the smallest, on top, and disk n, the biggest, at the bottom (Figure 8.7). The task at hand is to move the disks one by one from the first pole to the third pole, using the middle pole as a resting place, if necessary. There are only three rules to follow:

1. A disk cannot be moved unless it is the top disk on a pole. Only one disk can be moved at a time.
2. A disk must be moved from one pole to another pole directly. It cannot be set down some place else.
3. At any time, a bigger disk cannot be placed on top of a smaller disk.

To simplify our discussion, let us label the first pole A (source pole), the second pole B (the parking pole), and the third pole C (the target pole). If you have not seen the solution before, you might like to try a small example first, say, $n = 3$. It does not take long to figure out the following sequence.

move disk 1 from A to C
move disk 2 from A to B
move disk 1 from C to B
move disk 3 from A to C
move disk 1 from B to A
move disk 2 from B to C
move disk 1 from A to C

So it turns out that you need 7 moves for the case $n = 3$. As you get a feel of how to do three disks, you are tempted to do four disks, and so on. But you will soon find that there seems to be no rule to follow, and the problem becomes much harder with each additional disk. Fortunately, the puzzle becomes very easy if you think about it recursively.

Let us apply our recursion engine to this puzzle in order to generate a sequence of correct moves for the problem: *Move n disks from pole A to C through B.*

(i) Termination condition: If $n = 1$, then move disk 1 from A to C and return.

(ii) Subproblems: For $n > 1$, we shall do three smaller problems:

1. Move $n - 1$ disks from A to B through C
2. Move disk n from A to C
3. Move $n - 1$ disks from B to C through A

There are two smaller subproblems of the same kind, plus a trivial step.

(iii) Recombination of answers: This problem is solved after the subproblems are solved. No recombination is necessary.

Let's write down the recursive function for the solution. Two recursive calls and a move of disk n is all that it takes.

Algorithm `hanoi(n, a, b, c)`:
Input: Integer n (the number of disks), `a` (name of source pole), `b` (name of parking pole), `c` (name of target pole)
Output: Displays a sequence of moves

1. If $n = 1$, display "Move disk 1 from a to c" and return.
2. `hanoi(n-1, a, c, b)`
3. Display "Move disk n from a to c"
4. `hanoi(n-1, b, a, c)`

Each of steps 2 and 4 makes a recursive call. This looks almost too simple, doesn't it? But it works. To obtain a solution for 7 disks, say, we make the call

```
hanoi(7, 'A', 'B', 'C')
```

During the course of the solution, different poles are used as the source, middle and target poles. This is the reason why the `hanoi` function has the `a, b, c` parameters in addition to `n`, the number of disks to be moved at any stage. Figure 8.8 shows the three-step recursive solution for `n = 5`:

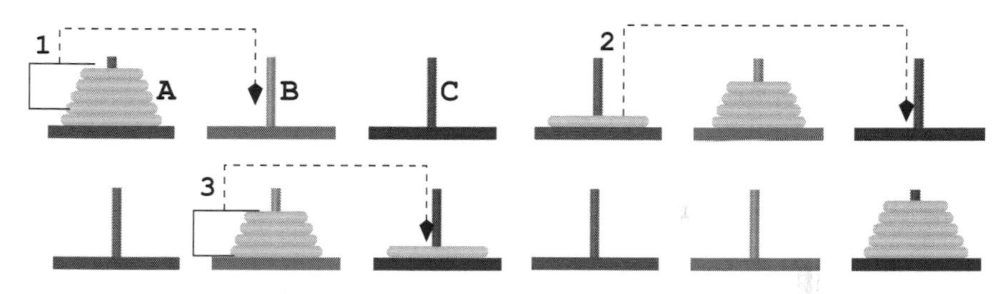

FIGURE 8.8 Tower of Hanoi (Five Disks)

1. Move 4 disks from pole `A` to pole `B` (`hanoi(4, 'A', 'C', 'B')`)
2. Move disk 5 from pole `A` to pole `C`
3. Move 4 disks from pole `B` to pole `C` (`hanoi(4, 'B', 'A', 'C')`)

Now, what is the number of moves needed for the `hanoi` algorithm for n disks? Let the move count be $mc(n)$. Then we have:

- If $n = 1$, then $mc(n) = 1$.
- If $n > 1$, then $mc(n) = 2 \times mc(n-1) + 1$

The above definition for $mc(n)$ is in the form of a *recurrence relation*, and it allows us to compute $mc(n)$ for any given n.

But, we can also seek a close-form formula for $mc(n)$.

$$
\begin{array}{ll}
n = 1 & mc(1) = 1 \\
n = 2 & mc(2) = 2 + 1 \\
n = 3 & mc(3) = 2^2 + 2 + 1 \\
n = 4 & mc(4) = 2^3 + 2^2 + 2 + 1 \\
\ldots &
\end{array}
$$

Generally:

$$mc(n) = 2^{n-1} + 2^{n-2} + ... + 2 + 1 = 2^n - 1$$

An interactive Tower of Hanoi game (**Demo:** `Hanoi`) can be found at the CT website. Because $2^n - 1$ moves will be needed for n disks, you should test the program only with small values of n. But the monks in Hanoi are not so fortunate, they have 200 heavy gold disks to move, and the sun may burn out before they are finished!

But the logic behind the *recursion engine* provides a problem solving strategy unparalleled by other methods. Many seemingly complicated problems can be solved easily with recursive thinking.

8.6 Eight Queens

The Bavarian chess player Max Bezzel formulated the Eight Queens problem in 1848. The task is to place eight queens on a chess board so that no queen can attack another on the board. As you may know, a queen can attack another piece on the same row, column, or diagonal. And the question is how many solutions are there.

It turns out that there are 12 basic solutions (Figure 8.9 shows one). Other solutions can be derived from these by board rotations or mirror reflections for a total of 92 distinct solutions.

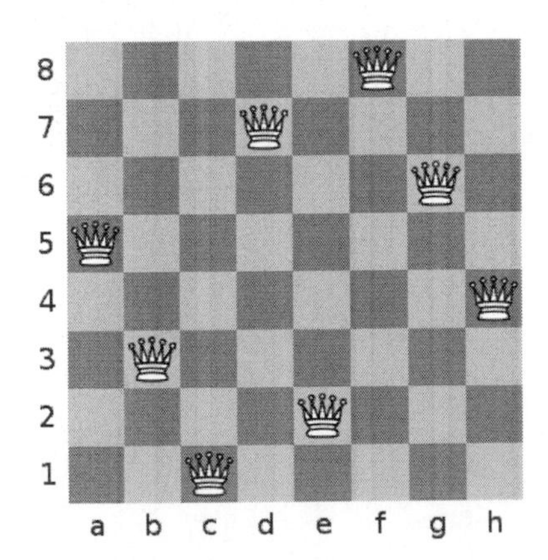

FIGURE 8.9 Eight Queens

We know a necessary condition for a solution is that each column and each row must contain one and only one queen. To satisfy this necessary condition, there are 8 possible column positions for row 1, 7 for row 2, and so on, for a total of 8! = 40320 queen placements. A brute-force way to find solutions is to check all 8! cases for diagonal attacks.

But we can do better than that by using a solution technique called *backtracking*. The idea is to place the queens, one at a time, making sure the next queen is placed in a nonattacking position in relation to previously placed queens. If the procedure finished placing all 8 queens, we have a solution. If it got stuck along the way, we backtrack to the previous queen and move it to its next possible position. If it has no next position, then we backtrack further. Compared to the brute-force method, backtracking examines far fewer cases.

To illustrate backtracking, let's look at a Four Queens problem. We begin by placing the first queen in the first column at board position (1,a) (Figure 8.10 left). Next we place our second queen in the second column at board

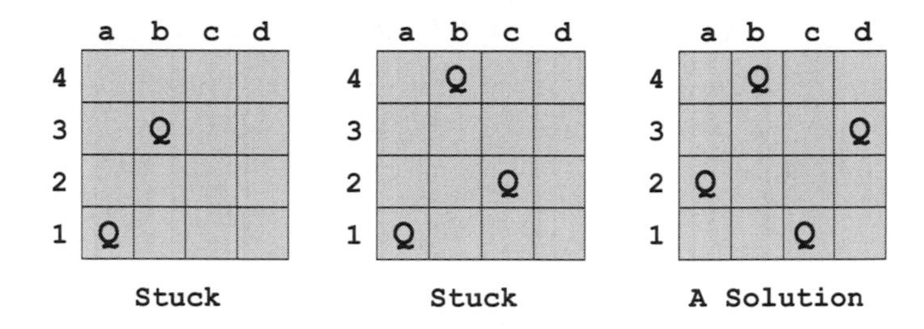

FIGURE 8.10 Four Queens Backtracking

position (3,b). We then found that there is no place for the third queen in the third column. We are stuck.

So, we go back and move the second queen to the next possible position in the column at board position (4,b). This allows us to place the third queen in the third column at position (2,c), only to find there is no position for the fourth queen (Figure 8.10 middle).

Again, we need to go back and change the position of a previous queen. It turns out that we have no next positions for the third or second queen. We are forced to move the first queen to its next possible position (2,a). From this position, proceeding in the same manner as before, we finally arrive at a solution (Figure 8.10 right). This is clearly faster than the brute-force approach. The efficiency of backtracking comes from abandoning further queen placements after getting stuck.

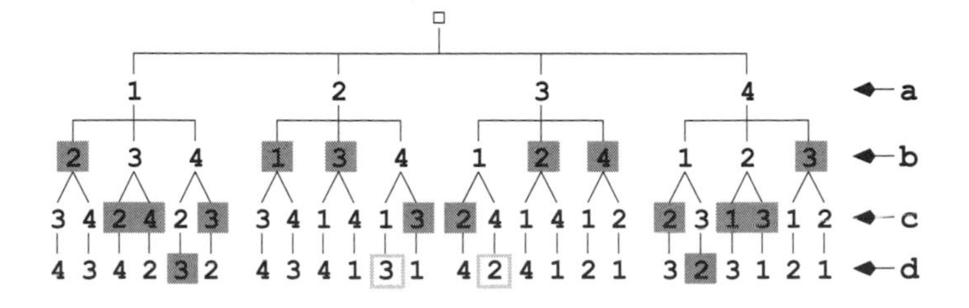

FIGURE 8.11 A Solution Tree

To illustrate the savings, let's look at the solution tree (CT: LEARN FROM TREES, Section 4.8.2) for the Four Queens problem (Figure 8.11), where the first level nodes give the row positions for the first queen (in column `a` on the board), the second level positions for the second queen (column `b`), and so on.

The solid shaded nodes are deadends. The paths leading to the two boxed nodes represent the only two solutions. You can see in Figure 8.11 how many tree branches are pruned by backtracking, resulting in significant computational savings.

Backtracking Implementation

Let's see how backtracking is applied to solve the Eight Queens problem by showing an algorithm for it. Our algorithm uses the following quantities:

- The size of the board is `N` by `N`; both rows and columns are numbered from 1 to `N`
- The integer array `qn`, with elements `qn[0]` through `qn[N]`
- The quantity `qn[c]` is the row position of the queen in column `c`, where $1 \leq c \leq N$

The backtracking algorithm attempts to position a queen in each successive column and is specified as a recursive function `queens`:

Algorithm `queens(r0, c)`:
Input: Integer `r0` (starting row), `c` (current column)
Effect: queens(1,1) displays all solutions for the `N` queens problem

1. If `(c < 1)`, then return (finished)
2. If `(c > N)`, then (found one solution)
    (a) Display `qn[1]` through `qn[N]` as the solution found
    (b) Call `queens(qn[c-1]+1, c-1)` (for more solutions)
    (c) Return
3. ```
   for (r=r0; r <= N; r=r+1)
   {  if ( safep(r, c) )
      {  set qn[c] = r;
         queens(1, c+1); (proceeds to next column)
         return;
      }
   }
   ```
4. (Backtrack) Call `queens(qn[c-1]+1, c-1)`
   ```

After setting N=8, the call queens(1, 1) produces all 92 solutions for the Eight Queens problem.

The function terminates (Step 1) if c is zero (backtracked to the left of column 1). Otherwise (Step 2), if c exceeds N (a queen in each column), it displays the solution and continues (Step 2b) to check for more solutions before returning.

When Step 3 is reached, the function tries to place a queen in column c at a valid row between r0 and N inclusive. After each successful row placement, it continues to the next column by calling queens(1, c+1). Finally (Step 4), having processed all rows between r0 and N, it backtracks to the previous column for more solutions.

The predicate **safep** checks the validity of position (r, c) for placing a queen and returns true or false.

```
safep(r, c)
{   for (y=1; y < c; y=y+1)
    {  if (qn[y] == r || abs(qn[y]-r)==abs(c - y))
          return 0;
    }
    return 1;
}
```

The function makes sure that the position (r, c) is not on an occupied row or diagonal by a queen in an earlier column.

See an interactive **Demo: Queens** at the CT site, where you can place queens on a board to find solutions interactively, and also see all 92 basic solutions for the Eight Queens problem.

8.7 General Backtracking

We used the Eight Queens problem to introduce the backtracking technique, which is generally applicable to solve problems by building a solution one element at a time. For a queens problem, we can place one queen at a time until all queens are placed. Other such problems include crossword and Sudoku puzzles. But backtracking is not just for games. It has wide applicability in solving practical problems, such as the *knapsack problem*, packing items of different weight or size into a container. The goal is to maximize the total dollar value, for example, of the packed items.

Similar to the queens problem, the knapsack problem is a type of *combinatorial optimization* problem where different combinations of items, satisfying given conditions known as *constraints*, are examined to optimize certain desired values. For the queens problem, we have the nonattacking constraint, and we want to find different ways to place the maximum number of queens on the board. For the knapsack problem, we want to find different ways to pack the given items in order to maximize the value of the packed items under certain measures.

8.8 Tree Traversals

We have talked about the organization of computer files in Section 4.8.2, where Figure 4.6 showed a typical file tree. A tree is a very useful way to organize hierarchical data (CT: LEARN FROM TREES, Section 4.8.2). Family trees are well-known. Internet domain names are also organized into a tree structure.

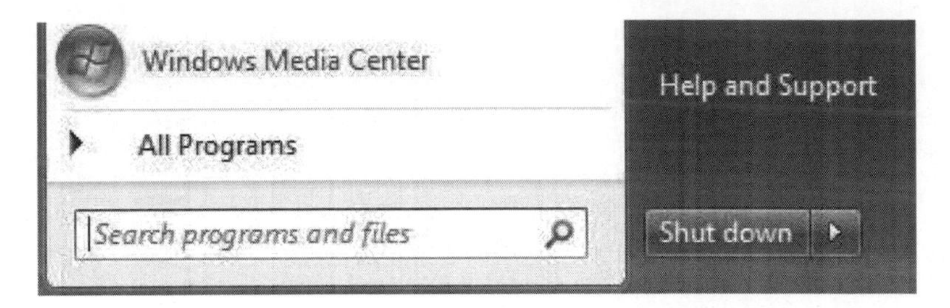

FIGURE 8.12 Windows File Search

On a computer, sometimes we need to search for a file because we have forgotten its folder location, or we are not sure of the file's precise name. Or, we want to find all files whose name contains a certain character string. Most operating systems provide a way for users to do such searches. It can be as easy as typing in a substring of the file name (Figure 8.12) and a computer program will look for files with matching names in the entire file tree. This is very convenient, indeed. In fact, you can also find files containing certain words. That can come in handy when you remember parts of the file contents but not the file name.

But, how can such search operations be performed? Well, we need a systematic way to visit each node on the file tree, known as a *tree traversal*. A file tree traversal enables a program to visit all files, following folders and subfolders, and to match file names or contents with user input.

The two most common tree traversal algorithms are *depth-first traversal* (dft) and *breadth-first traversal* (bft). With bft, we visit the root, then its child nodes, then grandchild nodes, and so on. With dft, we visit the root, then we **dft** the first child branch, **dft** the 2nd child branch, and so on. Figure 8.13 shows the node-visit order for a tree using bft and dft.

Implementation of the bft is straightforward. Because dft is defined recursively, we can use a recursive algorithm to implement it.

Algorithm `dft(nd)`:
Input: `nd` the starting node for dft traversal
Effect: visiting every node in the tree rooted at `nd` in dft order

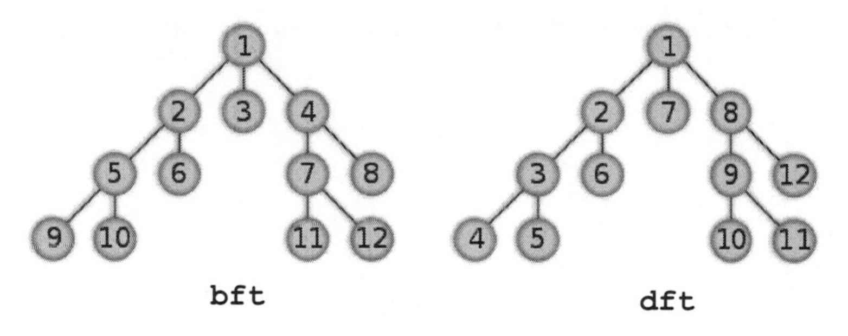

FIGURE 8.13 Tree Traversals

1. Visit `nd`
2. If `nd` is a leaf node (no children), then return
3. Otherwise, for each child node `c` of `nd`, from first child to last child, call `dft(c)`

Trace this algorithm when called on the root of the tree in Figure 8.13 and verify the dft order given in the figure.

> **CT: FORM TREE STRUCTURES**
>
> *Keep the tree structure in mind. It can be found everywhere and can be used to advantage in many situations.*

Let's take a fresh look at the solution tree (Figure 8.11) for the Four Queens problem in Section 8.6. We see now that the backtracking algorithm employed there is simply a dft of the solution tree while applying the nonattacking condition at each node. A solution is found whenever a valid last level leaf is reached.

Tree traversal is important in programming because tree structures are found in many varied situations. For example, markup languages, including HTML (Hypertext Markup Language) (Section 6.5), organize a documents into a tree structure of the top element containing data and child elements that may in turn contain data and other elements.

8.9 Complexity

The speed of modern computers adds a new dimension to problem solving, namely by brute force. We are no longer limited by our own speed or processing capabilities. Instead, we can ask the computer to examine all cases, or explore all possibilities, even when their numbers become quite large.

For example, we can solve the Eight Queens problem by checking each of 8! ways of placing the queens. The backtracking algorithm is a faster way to explore all of the solution tree, which has 8! branches. However, if the number of queens, n, increases much beyond 8, then the brute-force method, even with backtracking, will soon prove to be too slow. This is because the number of possible solutions $n!$ grows big very quickly as n increases.

CT: WEIGH SPEED VS. COMPLEXITY

Fast computers enable solutions by brute force. But, they are no match for rapidly growing problem complexities.

To get a feel of how fast $n!$ grows as n increases, let's consider a task of simply running a loop that does nothing for 20! iterations. Assuming a fast computer with a CPU (Central Processing Unit) clock rate of 10 GHz (10^{10} Hz), and each iteration takes just 1 clock cycle, we can compute how long the task will take:

$$20! = 2432902008176640000$$

$$\frac{20!}{10^{10}} = 243290200.817664 \text{ seconds}$$

$$\frac{243290200.817664}{24\times 60\times 60} = 2815.85880576 \text{ days}$$

That is more than 7.5 years! It will take much, much longer if each loop iteration actually does something.

The term *complexity* is used in computer science in two ways: (i) the inherent difficulty of computational problems, and (ii) the growth of time/space required by an algorithm as its input problem size increases.

Indeed, not all problems or algorithms are created equal. For example, binary search (Section 1.8) grows proportional to $log_2(n)$ in complexity, where n is the length of the sorted list. Finding the maximum/minimum value in an arbitrary list grows linearly with the list size. The bubble sort algorithm (Section 8.2.1) has complexity n^2, while the quicksort algorithm (Section 8.3.1) has an average complexity of $n \times log_2(n)$.

In computer science, the well-known *traveling salesman* problem asks this simple question: Given a set of cities and the distances between each pair of cities, what is the shortest route to visit each city and return to the starting city? Figure 8.14 shows an instance of this problem involving five cities. Such a problem has proven to be difficult when the number of cities grows. For n cities, there are $(n-1)!$ possible routes to check. This *combinatorial growth* is also seen in the the queens problem.

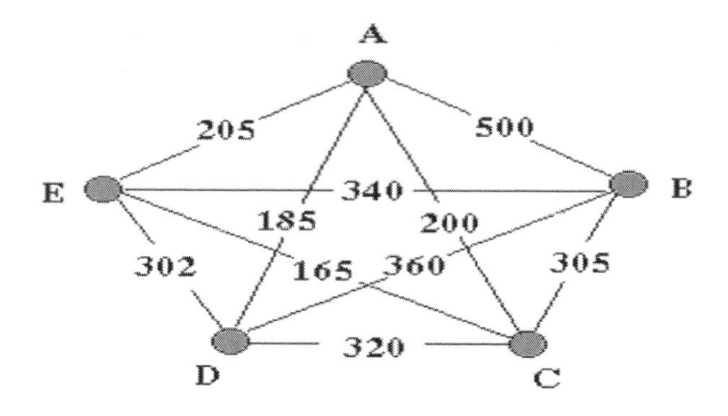

FIGURE 8.14 Traveling Salesman Problem

8.10 Heuristics

When faced with a high-complexity problem that quickly outstrips the computational powers of computers, what is a problem solver to do? Well, giving up is the last option. We must be resourceful and try our best to come up with something: a shortcut, an approximation, an oversimplification, or an experience-based rule of thumb. In other words, we will try *heuristics*.

In computer science, a *heuristic* is a technique to solve a problem more quickly or efficiently when brute-force or rigorous algorithmic methods are too expensive (practically impossible), or to get some solution instead of insisting on an exact or optimal one. This is often achieved by trading accuracy, precision, optimality, or completeness for computational feasibility.

In daily life, well-known examples of heuristics include stereotyping and profiling. Let's look at some examples in computing. We already know that the Eight Queens puzzle becomes too big for a slightly larger number of queens. But, we can apply the following heuristic to at least get a solution.

> **CT: DEVISE HEURISTICS**
>
> *Apply heuristics when problem complexity outstrips computational power.*

Here is a *queens heuristic*, where we form a list, *positions*, of row positions for queen placement based on the number of queens:

1. Let $N \geq 4$ be the number of queens. Let $even = (2, 4, 6, ...)$ be the list of even numbers, and $odd = (1, 3, 5, ...)$ be the list of odd numbers, less than or equal to N.

2. If $N \ mod \ 6 = 2$, then swap 1 and 3 in *odd* and move 5 to the end

3. If $N \ mod \ 6 = 3$, then move 2 to the end of *even*, and move 1 and 3 to the end of *odd*

4. *positions* = *even* followed by *odd*

Applying this heuristic to $N = 7$, we get $positions = (2, 4, 6, 1, 3, 5, 7)$. For $N = 8$, we get $positions = (2, 4, 6, 8, 3, 1, 7, 5)$. For $N = 20$, we get

$$positions = (2, 4, 6, 8, 10, 12, 14, 16, 18, 20, 3, 1, 7, 9, 11, 13, 15, 17, 19, 5),$$

and we did not wait for years!

For the traveling salesman problem (TSP), there are quite a few heuristics. A TSP *tour* is a round trip visiting each city once and returning to the origin city. The simplest and most intuitive is the *nearest city* heuristic, which calls for always traveling to the nearest new city from the current city. To find the nearest city at every step, we need to perform a total of

$$(n - 1) + (n - 2) + ... + 2$$

comparisons. The number is proportional to n^2.

The *greedy* heuristic sorts all, at most $\frac{n(n-1)}{2}$, segments between city pairs and form a sorted list, `L`. It then constructs a tour, usually shorter than the nearest city heuristic, by adding segments, one at a time, to the tour and removing them from `L`:

1. Add the shortest segment from `L` to the tour and remove it from `L`.

2. From `L`, add the shortest **valid** segment to the tour. A segment is invalid if it causes a city to be visited twice unless it's the nth segment.

3. Repeat step 2 until the tour is complete.

Note that segments of the tour may not be connected until construction is complete.

Another TSP heuristic constructs a complete tour by forming disjoint neighborhood subtours and merging them together. A subtour is a tour involving a subset of the cities. The heuristic starts with n subtours, each consisting of an individual city. Then it merges subtours following these rules:

- From all current subtours, pick the two closest subtours and merge them into a larger subtour.

- When merging two subtours, find the best way that minimizes the merged subtour.

Figure 8.15 illustrates the neighborhood tour heuristic for TSP.

Finding good algorithms to automate problem solution is at the center of modern computing. There are many good techniques including, brute-force iteration, chipping away, recursion, top-down divide and conquer, bottom-up

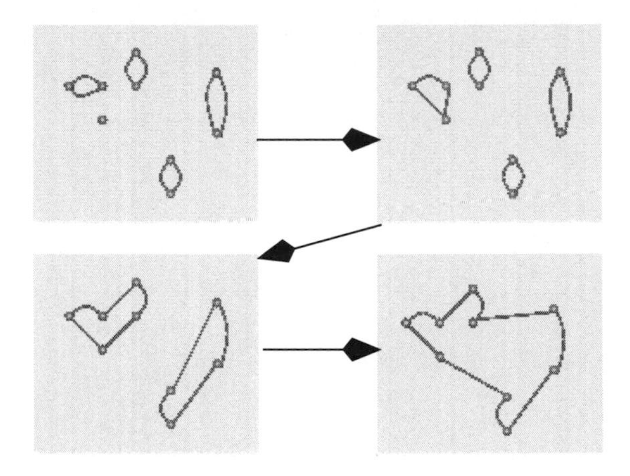

FIGURE 8.15 Neighborhood Tour Heuristic

building blocks, tree traversal, backtracking, and others. For problems that are too complex, the challenge is to come up with clever heuristics to at least get some results.

In general, problem solving involves these steps:

1. Precisely define the problem, task, or goal.

2. Consider different ways to tackle the problem, perform the task, or achieve the goal. Take into account the effectiveness, efficiency, and degree of complexity of the solution method.

3. Attempt to automate the solution by specifying it in an algorithmic form. This can make the solution more foolproof if not completely automatic.

While concepts and ideas in this chapter can help make you a better problem solver, don't forget to search the Web for answers before inventing a solution of your own.

Exercises

8.1. Solve the Magic Tray puzzle in Section 8.1.3. Hint: The configuration shown in Figure 8.3 is one step away from being done.

8.2. Consider the Magic Tray puzzle in Section 8.1.3. If we reach into two random pockets at each step, what is the chance of missing a particular pocket after n steps?

8.3. Refer to the improved `bubblesort` algorithm in Section 8.2.2. Show that the variable `go_on` is true at the end of a pass if and only if no call to `exchange` is made for any `i > 0` in the pass.

8.4. Consider the Eight Queens problem. Having positioned 7 queens, we are now trying to place the last queen. Is it true that there can be at most one valid position for the last queen? Why or why not?

8.5. Choice of the partition element (`pe`) is critical to the efficiency of quicksort. A strategy that works well in practice is to pick the first, middle, and last elements, and then choose the median value as the `pe`. Refine the `partition` function to implement this improvement (Section 8.3.1).

8.6. With your own words, explain the difference between an algorithm and a heuristic.

8.7. Study the Tower of Hanoi (Section 8.5) example and explain the power of recursion in your own way.

8.8. **Computize**: What are your own ideas about having a solution vs. having an algorithmic solution to a problem? Please explain at length.

8.9. **Computize**: Give two examples, not found in this book, of problems with a recursive solution.

8.10. **Computize**: Give examples of things with a recursive nature everywhere.

8.11. **Computize**: Estimate the height of the pile you get if you fold a piece of paper in half repeatedly 32 times.

8.12. **Group discussion topic**: *My understanding of recursion.*

8.13. **Group discussion topic**: *Thinking outside the box.*

8.14. **Group discussion topic**: *Solution by brute force.*

Chapter 9

Data Everywhere

The world runs on data. This was true long before the introduction of modern computers. The very first programmable digital computer, the ENIAC (Electronic Numerical Integrator And Computer, 1946), was designed to do fast number crunching to solve numerical problems, such as projectile trajectory in ballistics. Later, in 1951, the first UNIVAC I (UNIVersal Automatic Computer I) was accepted by the United States Census Bureau to process census data.

We now live in an interconnected world of information technology (IT), where all sorts of data are created, stored, processed, analyzed, and applied to great benefits. Data is everywhere: contact lists on our smartphones, events on calendars, academic records in schools, inventories in stores, addresses and ZIP codes in the postal system, bar codes for products, QR (Quick Response) codes, accounting in companies, transaction records in finance, police records in crime fighting, maps and GPS data in navigation systems, medical records in health care, measurements in weather prediction, streaming and broadcasting contents, ancestries in genealogy, DNA structures in biology, orbits of near-earth objects in astronomy. And the list goes on.

In fact, all the pages on the Web form an ever-expanding and changing set of data that has revolutionized our lives. Understanding digital data and their processing can significantly inform computational thinking.

When you come right down to it, modern computers execute instructions that manipulate bit patterns. They are natural numerical machines able to crunch numbers at great speed and precision. To turn this ability loose on data, we must find clever ways to represent, structure, and organize all kinds of data in numerical forms (bit patterns).

For a program, the instructions constitute a certain procedure or algorithm. Data made available to a program become its input. And data produced by a program become its output. Thus, a program can be regarded as a well-defined transformation of input data to output data.

> **CT: Garbage in, Garbage Out**
>
> *If the input is incorrect, then the output cannot be correct.*

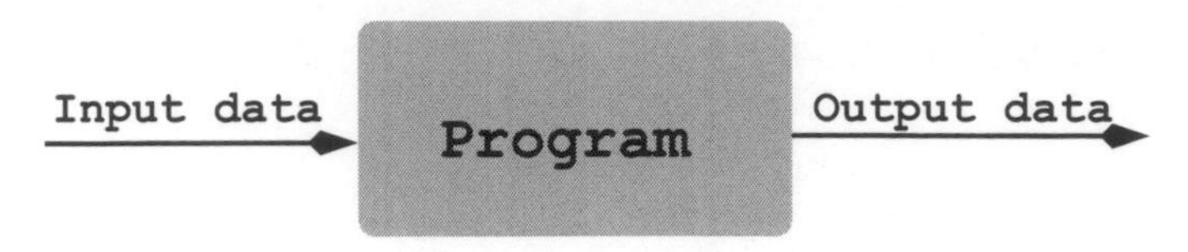

FIGURE 9.1 A Program

Obviously, no program can take incorrect input data and produce useful output data. We must make sure both the program and the input data are correct in order to obtain desirable output data.

In Chapter 2, we saw how bit patterns are used to represent numbers and characters. Now, we will see how more complicated data, such as images, audios, videos, employee records, and so on, are represented, processed, and stored.

In a real sense, programming consists of two main activities, designing and implementing algorithms and creating and manipulating data. The effectiveness and efficiency of a program depend on both its algorithms and its data structures/representations.

9.1 Digital Images

Modern displays, for HDTVs, computers, and smartphones, render images by illuminating individual *pixels* (picture elements) on the screen. The display screen consists of a rectangular array of pixels. Each row of the array contains the same number of pixels. A monitor with a denser pixel array gives higher *resolution* and the ability to display finer details. For example, a typical widescreen (16:9 aspect ratio) HD (high definition) monitor provides 1920 (column) × 1080 (row) of pixels.

Likewise, a digital camera also captures each image as an array of pixels. Typical resolutions include: 640 × 480, 1216 × 912 (1 Megapixels, MP), 1600 × 1200 (2 MP). And it goes much higher. Each individual pixel is specified by its color. Images so represented are known as *raster images.*

9.1.1 Representing Color

Let's take a look at how colors are represented as data. Of course, we have names for various colors: red, green, blue, yellow, cyan, magenta, and so on. But color names are limited and hard to use in computing. For on-screen display, the *RGB* (red-green-blue) *additive color model* is widely used. With RGB, a color is specified by the amounts of the three primary colors it contains. By adding the specified red, green, and blue lights at a pixel location, a certain color can be produced on the monitor screen.

Usually, three bytes (R, G, B) are used (Table 9.1). The value (0–255) of each byte specifies the red, green, or blue component of a color, allowing for $256^3 = 16,777,216$ different colors (*true color*)[1]. For example, `(255,0,0)` is full red, `(0,255,0)` is full green, and `(0,0,225)` is full blue. And `(0,0,0)` is black, `(255,255,255)` is white. Often, each RGB byte is written down as two hex digits. For example, `color="#FFFFFF"` sets the foreground color to white in a webpage (Section 6.6).

TABLE 9.1 24-bit RGB Color Representation

	R								G								B							
Black	0	0	0	0	0	0	0	0	0	0	0	0	0	0	0	0	0	0	0	0	0	0	0	0
White	1	1	1	1	1	1	1	1	1	1	1	1	1	1	1	1	1	1	1	1	1	1	1	1
Red	1	1	1	1	1	1	1	1	0	0	0	0	0	0	0	0	0	0	0	0	0	0	0	0
Green	0	0	0	0	0	0	0	0	1	1	1	1	1	1	1	1	0	0	0	0	0	0	0	0
Blue	0	0	0	0	0	0	0	0	0	0	0	0	0	0	0	0	1	1	1	1	1	1	1	1

For printing hard copies, the *CMYK* (cyan-magenta-yellow-black) *subtractive color model* is usually used. Ink used for printing absorbs specific colors while being transparent to other colors. Combining different inks takes out unwanted colors (from a white background) to reflect a desired color.

- Cyan ink—Absorbs red (no reflection of red from white background)
- Magenta ink—Absorbs green
- Yellow ink—Absorbs blue
- Black ink—Reduces light overall reflection to produce gray and black

When cyan, magenta, and yellow inks are mixed together at their full density, in theory, we should get black. But, this is usually not true in practice because the white paper can reflect some light through the semitransparent inks. Black ink is used to add the necessary ink density to produce darker colors. Only black ink is needed to produce black for printing. When an image represented in RGB is printed, it is usually automatically converted to CMYK, where the value of each component ranges from 0 (none) to 1 (full strength). Table 9.2 lists a few examples.

A color may also be given an *alpha* value indicating its *opacity*, a property that determines how much or how little of its background shows through. For $alpha = 0$, a color is completely transparent (invisible) and lets its background show through completely. For $alpha = 1$, a color is totally opaque and will not let any of its background show through. Transparency increases and opacity decreases as the *alpha* value varies from 1 to 0.

[1]The human eye can distinguish about 10 million different colors.

TABLE 9.2 RGB to CMYK Conversion

Color	(r,g,b)	(c,m,y,k)
Red	(255,0,0)	(0,1,1,0)
Green	(0,255,0)	(1,0,1,0)
Blue	(0,0,255)	(1,1,0,0)
Cyan	(0,255,255)	(1,0,0,0)
Magenta	(255,0,255)	(0,1,0,0)
Yellow	(255,255,0)	(0,0,1,0)
Black	(0,0,0)	(0,0,0,1)

9.2 Raster Image Encoding

Data for raster images specify colors of a grid of pixels. The finer the grid, the better the resolution. The more bits per pixel, the truer the color. The total memory size required for this *raw image data* would be (number of pixels) $\times$ (number of bits per pixel). For example, a 13 MP image may take $(4208 \times 3120) \times 24 = 315095040$ bits or about 37.56 MB of memory.

9.2.1 Raster Image Formats

A raster image usually does not need to record each and every pixel individually. *Compression methods* can significantly reduce the image data size. There are a number of standard image compression formats in common usage. They employ different *compression and decompression* (codec) schemes (Section 9.8). The whole codec process may (*lossy compression*) or may not (*lossless compression*) lose some details or quality contained in the raw image data.

Widely used image compression formats include:

- Graphics Interchange Format (GIF)—A raster format suitable for icons, logos, cartoons, and line drawings. GIF images can have up to 256 colors (8-bit). GIF files use the `.gif` file name suffix.

- Portable Network Graphics (PNG) format—A format designed to replace GIF. PNG really has three main advantages over GIF: alpha channels (color transparency), gamma correction (cross-platform control of image brightness;), and two-dimensional interlacing (a method of progressive display). PNG files use the `.png` file name suffix.

- Joint Photographic Experts Group (JPEG) format—A lossy raster format widely used for color and black-and-white photographs that usually have continuously changing color tones. JPEG images can store up to 24-bit RGB color and achieve good compression ratio. But it does not support transparency. JPEG files use the `.jpg` or `.jpeg` file name suffix.

- Tagged Image File Format (TIFF)—A lossless format widely used primarily in printing, often using the CMYK color space. It also supports transparency. Images created by a scanner or digital camera are usually stored in TIFF. TIFF files use the `.tif` file name suffix.

Even with JPEG compression, a typical digital photograph takes about 2.5 MB storage. Such a large file is more expensive to transmit, store, and process. And the large size is usually not necessary for most purposes.

CT: Small Is Beautiful

Reduce the camera resolution when taking pictures to share online.

People often take pictures with their smartphones or digital cameras to be shared with others by email or texting. In such cases, there is no need for the full resolution, 8–20 MP or better available on most devices, which can result in large picture files. Instead, the camera image resolution should be set to 2 MP or less. In fact, a 640×480 photo (about 0.3 MP), at less than 300 KB, can do well for most email, texting, and on-Web sharing.

The higher MP settings should be used only to take pictures to be printed or enlarged.

9.2.2 Vector Graphics

Not all pictures are represented as raster images. *Vector graphics* uses coordinates and formulas to record points, lines, curves and other geometric objects to represent an image.

Vector image encoding avoids representing every pixel. Hence, vector images are usually smaller and easier to scale up or down. Also, geometric information can be combined with raster information to represent a complete image. Display software must understand the geometric information to render vector graphics files.

9.2.3 Scalable Vector Graphics

The Web supports vector graphics well. For example, HTML5 allows Scalable Vector Graphics (SVG), and JavaScript provides vector graphics programming. Also, the widely used Adobe Flash™ supports vector graphics.

As an example, Figure 9.2 shows the company logo of `webtong.com`, which simply consists of a red triangle and two blue triangles. The box and coordinates shown are not part of the logo.

The SVG code for this logo is

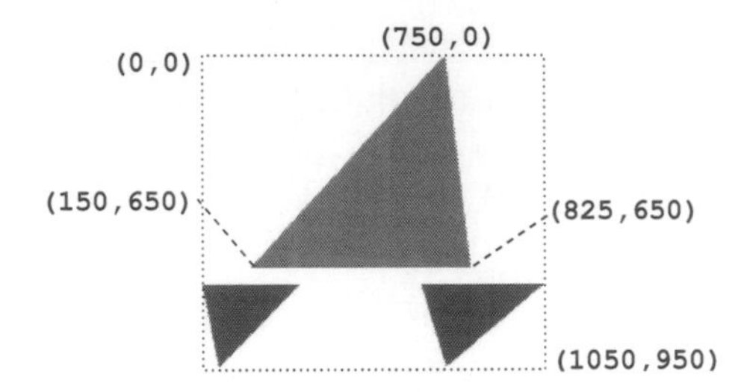

FIGURE 9.2 Sample SVG Graphic

```
<svg id='wtlogo' width='210' height='190'
                 viewBox='0 0 1050 950'>                  (A)
   <path d="M750,0 L150,650 825,650, z" fill="red" />     (B)
   <path d="M675,700 L1050,700 750,950, z" fill="blue" />
   <path d="M0,700 L300,700 50,950, z" fill="blue" />
</svg>
```

where the the triangles are drawn in a rectangular display area whose upper left corner is at `(0,0)` and lower right corner at `(1050,950)` (line `A`). The red triangle starts at `(750,0)`, to `(150,650)`, then to `(825,650)`, and finally returns to the starting point (line `B`). A 210×190 PNG image of the same logo is 3503 bytes, whereas the SVG version is only 274 bytes, independent of image size. See **Demo:** `SVGDemo` and **Demo:** `WebtongLogo` for an interactive experience with SVG graphics.

9.3 Audio and Video

9.3.1 Digital Audio

Audio and video form a big part of the Web and Internet. Technically, audio refers to sound within the human hearing range. An audio signal is naturally a continuous oscillating wave representing amplitudes. Analog audio must be digitized for processing and playback on computing devices.

An analog audio signal is digitized by *sampling* and *quantization*. Sound is caused by vibration. A sound wave represents the amplitude (volume) and frequency (pitch) of sound. The continuous sound wave is sampled at regular time intervals, and the amplitude value at each sampling point is quantized to the nearest discrete level (Figure 9.3). The resulting data are stored in binary format as a digital audio file. The higher the sample rate and the greater the bit depth (number of quantization levels), the higher the sound fidelity and the larger the file size. The same sampling and quantization process takes place in digitizing an image into an array of pixels.

Let F be the highest frequency of an audio signal. The sampling rate must be at least $2F$ to represent the signal well. This is the so-called *sampling*

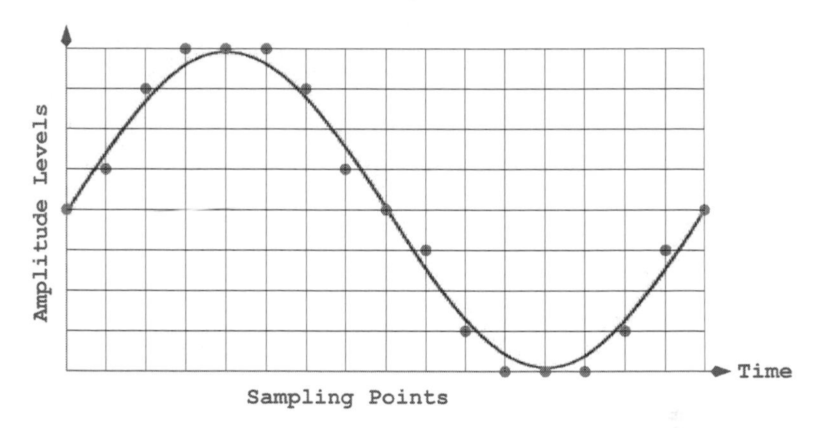

FIGURE 9.3 Sampling and Quantization

theorem. Human hearing is limited to a range of 20 Hz to 20K Hz (cycles per second). Thus, the CD-quality sampling rate is often 44.1K Hz. Human speech is limited from 20 Hz to 3K Hz. An 8K Hz sampling frequency is high enough for telephony-quality audio.

9.3.2 Audio Encoding Formats

Advances in digital audio bring increasingly sophisticated compression schemes to reduce the size of audio files while preserving sound quality. For example, the widely used MP3 is the audio compression standard ISO-MPEG Audio Layer-3 (IS 11172-3 and IS 13818-3).

In 1987, the Fraunhofer Institute (Germany), in cooperation with the University of Erlangen, devised an audio compression algorithm based on *perceptual audio*, sounds that can be perceived by the human ear. Basically, MP3 compression eliminates sound data beyond human hearing. By exploiting stereo effects (data duplication between the stereo channels) and by limiting the audio bandwidth, audio files can be further compressed. The effort resulted in the MP3 standard. For stereo sound, a CD requires 1.4 Mbps (megabits per second). MP3 achieves CD-quality stereo at 112–128 Kbps, near CD-quality stereo at 96 Kbps, and FM radio-quality stereo at 56–64 Kbps. In all international listening tests, MPEG Layer-3 impressively proved its superior performance, maintaining the original sound quality at a data rate of around 64 Kbps per audio channel.

MP3 is part of the MPEG audio/video compression standards. MPEG is the Moving Pictures Experts Group, under the joint sponsorship of the International Organization for Standardization (ISO) and the International Electro-Technical Commission (IEC). MPEG works on standards for the encoding of moving pictures and audio. See the MPEG home page (`mpeg.org`) for further information.

More recently, Ogg Vorbis (`xiph.org`) offers a fully open, nonproprietary, patent-and-royalty-free, general-purpose audio format that is preferred by many over MP3. Table 9.3 lists common audio formats.

TABLE 9.3 Audio Formats

Filename Suffix, Format	MIME Type	Origin
`aif(f)`, AIFF AIFC	`audio/x-aiff`	Apple, SGI
`mid`, MIDI	`audio/midi`	For musical instruments
`mp3`	`audio/mpeg`	MPEG standard
`ra` or `rm`, Real Audio	`audio/x-realaudio`	Real Networks
`amr`	`audio/amr`	Mobile phone audio, 3GPP
`ogg` or `oga`, Vorbis	`audio/ogg`	Open source, Xiph.Org
`wav`, WAVE	`audio/x-wav`	Microsoft
`wma`, Windows Media Audio	`audio/x-wma`	Microsoft

Actually, Ogg is a *media container* format (Section 9.4.1) rather than a single codec format.

Format converters are freely available to rewrite audio/video files from one format to another.

9.4 Digital Video

A video is a sequence of images displayed in rapid succession that is usually also played in synchrony with a sound stream. For smooth motion, a sufficient *frame rate*, about 30 frames per second (fps), is needed. There are many digital video formats for different purposes, including DVD, Blu-ray™, HDTV, DVCPRO/HD, DVCAM/HDCAM, and so on.

A video file usually supplies video tracks, audio tracks, and metadata. Tracks in a video file can also be organized into chapters that can be accessed and played directly. Such files are known as *video containers*, and they follow well-designed *container formats*, which govern the internal organization of a video file.

9.4.1 Video Containers

Widely used video container formats include:

- MPEG 4—A suite of standard audio and video compression formats from the Moving Pictures Experts Group (content type `video/mp4`; file suffix `mp4`, `mpg4`).
- Mobile phone video—A 3GPP-defined container format used for video for mobile phones (content type `video/3gpp`; file suffix `3pg`).

- Flash Video—A proprietary format used by Adobe Flash Player (content type `video/x-flv`; file suffix `flv`).
- Ogg—A free and completely open standard from the Xiph.Org Foundation. The Ogg video container is called Theora (content type `video/ogg`; file suffix `ogv`). The Ogg audio container is called Vorbis (content type `audio/ogg`; file suffix `ogg`).
- WebM—A new standard developed by `webmproject.org` and announced in mid-2010. The format, free and completely open, is used exclusively with the VP8 video codec and Vorbis audio codec. It is supported by, among others, Adobe, Google, Mozilla, and Opera. Google offers a WebM plug-in component for IE9. WebM is a new media container designed for the Web and is recommended for use with the `audio` and `video` elements of HTML.
- AVI—Audio Video Interleaved format from Microsoft (content type `video/x-msvideo`; file suffix `avi`).
- RealVideo—An audio and video format by Real Networks. For historical reasons, it uses the seemingly incorrect content type `audio/x-pn-realaudio` and the file suffix `rm`.

9.4.2 Video Codecs

The video and audio tracks in a container are delivered with well-established compression methods. A video player must decompress the tracks before playing the data. Many compression-decompression algorithms exist. Generally speaking, video compression uses various ways to eliminate redundant data within one frame and between frames.

The most important video codecs for the Web include H.264, Theora, and VP8.

- H.264—A widely used standard from MPEG providing different *encoding profiles* to suit devices from smartphones to high-powered desktops. Also known as *MPEG-4 Advanced Video Coding*, the standard enjoys good hardware and software support.
- HEVC—High Efficiency Video Coding, jointly developed by ISO and MPEG, HEVC can double the compression ratio of H.264/MPEG-4 AVC while maintaining the same level of video quality.
- Theora—A completely free standard from Xiph.Org Foundation providing a modern codec that can be embedded in any container. But, it is usually delivered in an Ogg container. Theora is supported on all major Linux distributions, Windows, and Mac OS X.

- VP8—A very modern and efficient standard from On2 (part of Google) giving everyone an open source and royalty-free codec for the Web. VP8 is usually delivered inside the WebM container.
- WMV—Windows Media Video, a family of proprietary video codecs, including WMV 7, WMV 8, WMV 9, and VC-1.

9.5 Format of Data and Files

Data and application programs are usually intimately related. Often, an application receives input data, processes the data, and produces output data. Such data are typically saved in files for later use or sharing with others. There are many different types of files to satisfy various kinds of application and intended purposes.

Typically, the first thing when processing a file is to determine its *file type*. Only the correct application programs can process files of certain types. Imagine what would happen if a music-playing application is given a spreadsheet file to process! The sound must be awful, if it works at all.

Files fall in two broad categories: *text file* and *binary file*. The former contains only characters in ASCII or UNICODE (Section 2.6), the latter a sequence of arbitrary bytes. PNG images, MP3 songs, and compiled programs, for example, are stored in binary files. Emails, Web pages, and Shell scripts, in addition to plaintext notes, are in text files. Text files have the advantage of being easily readable by humans and modifiable using simple text editors. Binary files require specific applications but are usually more compact and efficient.

CT: INTERPRETING DATA

Each file or content type defines its own data format, structure, and encoding.

Without knowing the type, data cannot be interpreted correctly. Ways to indicate the file/data type have been evolving together with advancements in operating systems, networking, and the Internet. In addition to *file name suffix* and standard MIME content types, as we have seen in Section 4.8, *in-file metadata* can also be used to specify the file type as well as other properties of a file. Such files follow a well-defined *file format* that governs the file's internal organization such as the *file header* structure, any *magic number*, and other metadata.

Table 9.4 lists some common file formats associated with well-known applications. A file format can be proprietary (`.psd`, for example), open but may require a license (`.doc`, `.xls`, `.ppt`, and `.mp3`, for example), or open and free (`.html` and LibreOffice formats, for example).

TABLE 9.4 Application and Data

Application	File Types	Open & Free
Microsoft Word	.doc, .docx, .docm	
Microsoft Excel	.xls, .xlsx, .xlsm	
Microsoft PowerPoint	.ppt, .pptx, .pps, .ppsx	
LibreOffice Word Processing	.odt, .fodt	yes
LibreOffice Spreadsheet	.ods, .fods	yes
LibreOffice Presentation	.odp, .fodp	yes
Web browser	.htm, .html	yes
Email readers	.eml, .emlx	yes
Adobe Photoshop	.psd, .psb	
Gnu Image Manipulation Program	.xcf	yes
Adobe Acrobat (Acroread)	.pdf	
LaTeX	.tex, .sty, .toc, .idx	yes

CT: Data Is Application Dependent

Data and files are usually represented differently for different applications.

The PDF file type is worth noting. It has become the defacto standard for sharing documents online across all platforms. Putting a PDF file online or in an email is the best way to avoid printing and/or faxing. Besides, contents of a PDF file can be searched automatically.

CT: Save Trees with PDF

Never print or fax a document if you can save/send it as PDF.

Applications use their own *native* file/data formats for effectiveness, efficiency, and sometimes proprietary reasons. Use the `file->save` option to save any changes and overwrite the original version, use `file->save as` to save it under a different name, use `file->export` to save it into some other format. Application can usually also work with non-native format files by first *importing* them, converting to them to the native format.

Programs are often available to convert files between comparable formats. Freely available audio and video converters are examples.

9.6 Data Sharing

Using data from others is one thing, supplying data you created to others is another. It is not unusual for an individual to want to share pictures, music, audio, video, and other types of documents either publicly or privately. Here are a number of ways to do it.

- Email—Send the files as email attachments. Beware that email systems have limitations to the size of attachments. Generally, if they total under 10MB, you should be OK.
- Website—Upload the files to your own website and send their URLs to others to share. Beware that Web access is public unless the files are placed in a protected realm (requiring login).
- The Cloud—Don't have a Website? Upload the files to the Cloud to share, using such online services as Dropbox™, Facebook™, Google Drive™, Picassa™ (pictures), YouTube™ (videos), and Spotify™ (music).
- FTP—Upload files to an FTP server to share with others.
- Bittorrent—Prepare a *torrent* for the files and folders you wish to share via Bittorrent. A torrent is a file that contains meta data about the contents to be shared, how they are cut into pieces, and other useful information. Upload the contents and the torrent to a Bittorrent site for others to download.
- Physical media—Store the files on a Flash drive or a CD/DVD and send it to someone.

When you have a group of files to share, consider using compression formats, such as `.zip`, `.tgz`, or `.rar` (Section 9.8).

9.7 Document Markup

Treating textual documents as data items is an important aspect of information technology. An effective way to organize complicated textual documents for easy automated processing is to delineate the role or meaning of pieces and parts contained inside the document. To make such delineations is to *markup* the document. For example, we have seen how HTML tags are used to markup Web pages (Section 6.5). The HTML tags are key for Web browsers and JavaScript programs to process, display, and otherwise manipulate Web pages and their constituent parts.

HTML is a *markup language* for organizing Web pages. But what about other documents, such as tax returns, employee records, product catalogs, and so on? Well, XML, the *eXtensible Markup Language*, is a technology that provides a way for anyone to invent and design a set of markup tags for any desired type of document.

Markup makes documents easy to communicate among heterogeneous platforms and efficient to process by programs.

9.7.1 What Is XML?

XML is not a markup language with a set of tags specific to structuring a certain kind of document. In fact, XML defines no tags at all. Instead, XML is a set of rules by which you can invent your own set of tags to organize and structure any specific type of textual documents you wish. Each organizational unit, usually bracketed by begin and end tags, is called an *element*. Thus, XML is a way to define your own markup language. Each XML-defined markup language is known as an *XML application*. Documents so marked up are called XML documents. XML is also a Web standard and there are many tools, mostly free, to create, process, and display XML documents.

SVG (Section 9.2.3) is just one of a large number of XML applications. For example, RSS (Rich Site Summary) and Atom (Atom Syndication Format) are standard Web news syndication formats based on XML (Section 9.7.3). Figure 9.4 shows the two familiar news syndication icons.

The list of publicly used XML applications is long (see the wikipedia). And then there are nonpublic markups used within organizations and even by individuals. XML makes it possible and easy for anyone to create a new markup language (CT: MARK IT UP, Section 6.5).

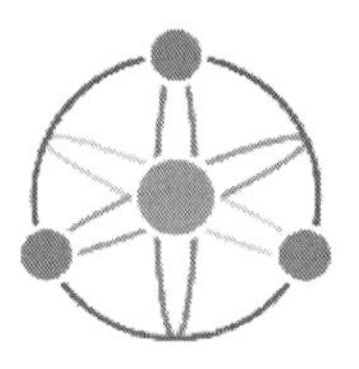

FIGURE 9.4 News Syndication Icons

9.7.2 XML Document Format

As shown in Figure 9.5, an XML document begins with processing instructions, followed by a *root element* that may contain other elements. The elements are defined by individual XML applications following rules given by XML.

```
Processing Instructions
<root_element>
   child elements ...
</root_element>
```

FIGURE 9.5 An XML Document

For example, here is a simple XML document:

```
<?xml version="1.0" encoding="UTF-8"?>          (1)
<address>                                       (2)
    <street>432 Main Street</street>
    <city>Arbville</city>
    <state>Ohio</state>
    <zip>02437</zip>
</address>
```

The `<?xml` processing instruction (line `1`) is required as the first line of any XML document. It provides the XML version and the character encoding used for the document. Other processing instructions follow the first line and can supply style sheets, definitions for the particular type of XML document, and other meta information for the document. Coming after the processing instructions is the root element that encloses the document content. The root element in this example is the `address` element (line `2`), which constitutes the remainder of the XML document.

In Section 9.2.3, we have seen how SVG (an XML application) is used to markup a vector-based image.

9.7.3 XML for News Syndication

XML is used widely in practice. Perhaps the best-known example is news syndication on the Web. The Web standard RSS is an XML application for *Web feeds*, and frequently updated content, such as blog entries and news headlines. An RSS document can supply an abstract, author, organization, publishing dates, and a link to the full story or article. Using a Web browser or an RSS reader, end users can *subscribe* to RSS feeds and easily see the changing headlines.

Here is a simple RSS example, the `books.rss` file:

```
<?xml version="1.0" encoding="ISO-8859-1" ?>
<rss version="2.0"><channel>
 <title>Sofpower Book Sites</title>
 <link>http://sofpower.com</link>
 <description>Companion websites for textbooks</description>
 <item>
   <title>Mastering Linux</title>
   <link>http://ml.sofpower.com</link>
   <description>Resources for "Mastering Linux",
   a highly recommended Linux book (2010)</description>
 </item>
 <item>
   <title>Dynamic Web Programming and HTML5</title>
   <link>http://dwp.sofpower.com</link>
   <description>Resources for "Dynamic Web Programming and
   HTML5", an in-depth Web programming book</description>
 </item>
</channel></rss>
```

The file lists two information/news items, about books published recently by the author, from the sofpower.com (channel) each linking to a website.

Atom is another XML-based news syndication standard from the IETF (Internet Engineering Task Force).

Web browsers and news reader applications understand RSS and Atom formats. Clicking on a link leading to an RSS or Atom file causes a Web browser to ask you if you wish to subscribe to the news feed. If you do subscribe, your browser will bookmark the feed and monitor any new headlines from the feed.

News feed links are usually marked by an orange-colored icon. See `cnn.com/services/rss` for RSS feeds from CNN, or

`hosted2.ap.org/APDEFAULT/APNewsFeeds`

for Atom feeds from AP.

The first step in processing the contents of a document is to discover its structure and the values for its various parts. This requires *parsing*, which may be difficult or even impossible for arbitrary documents. Markup tags change all that by making the document structure explicit.

CT: MARKUP FOR INTEROPERABILITY

Use XML to make data in your documents easily sharable and usable across heterogeneous computing systems.

9.8 Data Compression

It takes space to store data and time to transmit data. *Data compression* techniques can help reduce these requirements. With data compression (Figure 9.6), a *message* (data file or stream) is *deflated* (reduced in size) before storage or transmission. A deflated message can later be *inflated* to recover the original message, either exactly (in *lossless compression*) or approximately (in *lossy compression*). Some compression algorithms, such as those used in JPEG, GIF, and PNG (Section 9.2.1), are application specific. Others are more general and will work for nearly all message types.

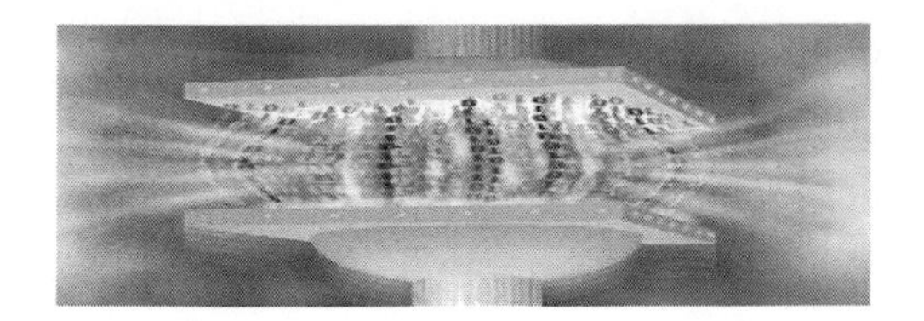

FIGURE 9.6 Data Compression

A compression algorithm takes advantage of the nature of the target messages to reduce their size. Consider black-and-white, two-tone images, those from FAX machines, for example. The pixels can be represented as a sequence of pixels; each is either white or black. But we can compress such a message using a run-length scheme that reduces the message into a sequence of integers indicating the number of consecutive white or black pixels. This can work well for ordinary document pages that are mostly white with a few black pixels here and there. It won't work at all if the whole page is filled with tiny black dots. For such a page, the run-length encoded message can become much larger. Compared to run-length, the actual algorithms used for JPEG, PNG, MP3, and so on are much more sophisticated.

Generally, a message, consisting of one or more files, can often be treated as a stream of characters or bytes. The LZ77 (by Abraham Lempel and Jacob Ziv, 1977) algorithm and the Huffman coding (David A. Huffman, 1952) for message deflation are very effective and employed by many popular compression applications (Table 9.5). It depends on the particular message, but these compression programs can often deflate a message to a third of its original size, or better.

While both data compression and data encryption (Section 7.4) transforms data into forms that must be unwound before being used, they serve very different purposes.

TABLE 9.5 Lossless Data Compression Applications

Application	Compressed File Suffix
ZIP	`.zip`
Bzip2	`.bz2, .tbz2`
Bzip	`.bz, .tbz`
Gnu Zip	`.gzip, .gz, .tgz`
WinRAR	`.rar`

CT: Compression Is Not Encryption

Compression reduces data size. Encryption secures data content.

An encrypted message cannot be decrypted unless the correct password is given. A compressed message, although not immediately readable, can be decompressed by anyone using readily available tools. No password is required.

9.8.1 LZ Deflation

LZ77 deflation replaces each repeated string in the source message by a reference to an earlier occurrence in the message. Such a reference is in the form of a length-distance pair

(len, d),

and it means "a sequence of *len* bytes, same as that *d* bytes earlier in the original message." This is effective when the original message is lengthy and contains many repeated sequences of bytes. Such is the case with HTML, XML, and plaintext documents. Here are some examples:

Source message = `teamwork, the way a team works`
encoded message = `teamwork, the way a (4,20) (4,21)s`

Source message = `bgfgfgfgbggb`
encoded message = `bgf(5,2)(2,8)gb`

As you can see, an LZ77 deflated message basically consists of references and *literals*, single characters or bytes not represented by references. The literals and numbers in references can usually be further deflated using Huffman code.

9.8.2 Huffman Code

In a particular message or type of messages, the frequencies of different symbols are certainly not equal. Some symbols happen more often than others and certain symbols hardly at all. For example, in a textual document, you would expect the letters `r`, `s`, `t`, as well as the vowels, to be more frequent than, say, `x`, `y`, and `z`; and characters such as ^ hardly at all. We can take advantage of such frequency differences to save space in representing characters. The basic idea of Huffman code is simple. Instead of using the same number of bits to represent each character (as in ASCII and UNICODE), we can use fewer bits for frequent characters and more bits for other characters.

The same goes for numbers. Instead of using, say, a 32-bit integer representation, we can use just a few bits for high-frequency numbers. In practice, you'll find small numbers much more often.

To Huffman encode a message, we first find the frequencies of symbols to be encoded. Based on the frequencies, we can build a *Huffman binary tree* (CT: FORM TREE STRUCTURES, Section 8.8), which defines how these symbols will be encoded into bit strings of various lengths. The Huffman tree is stored as part of the deflated message because it is needed for inflation.

As an example, let's apply Huffman coding to deflate a love poem by Nima Akbari (**Demo:** `HuffCode`), which begins

> You're my man, my mighty king,
> And I'm the jewel in your crown,
> You're the sun so hot and bright,
> I'm your light-rays shining down,
> . . .

The plaintext file for the whole poem contains 545 bytes, including spaces and line breaks. The character frequencies are shown here in increasing frequency.

```
N 1     - 1     . 1     W 1     p 1     j 2
v 3     A 5     f 5     k 5     c 6     b 7
Y 7     I 8     w 8     l 11    g 15    ' 15
s 15    d 17    y 17    , 17    m 18    t 20
u 20   CR 21    a 22    h 24    i 26    o 30
r 30    n 38    e 39   SP 89
```

Note characters include uppercase and lowercase letters and punctuation marks, including SPACE (`SP`), and RETURN (`CR`).

A list, `fq`, of character-frequency pairs, ordered in increasing frequency, such as the above, is used in a simple recursive algorithm to build a *Huffman tree*. Basically, the algorithm builds a binary tree from the bottom up, by creating tree nodes with characters having the least frequencies first. Each recursive call `huffmanTree(fq)` passes a list `fq` that is smaller in size.

Algorithm `huffmanTree(fq)`:
Input: Character-frequency list `fq`, in increasing frequency order
Output: Returns a Huffman binary tree

1. If `fq` is empty, then return an empty binary tree node
2. If `fq` has only one entry and that entry is a character, then return a leaf node with that character its child
3. If `fq` has only one entry and that entry is a tree node, (*node*, *frequency*), then return *node*
4. Remove the lowest frequency two entries *x* and *y* from `fq`; create a new binary node, `node`, with *x* and *y* as children
5. Set `fr` = frequency(x) + frequency(y); insert a new entry `(node, fr)` into `fq`, making sure that the list `fq` stays ordered
6. Return `huffmanTree(fq)`

In the algorithm, the word "character" means a single byte. The constructed Huffman tree is used to generate the Huffman code for each character/byte in the message to be deflated. Figure 9.7 shows a Huffman tree constructed this way. Starting from the tree root and using 0 for the left branch and 1 for the

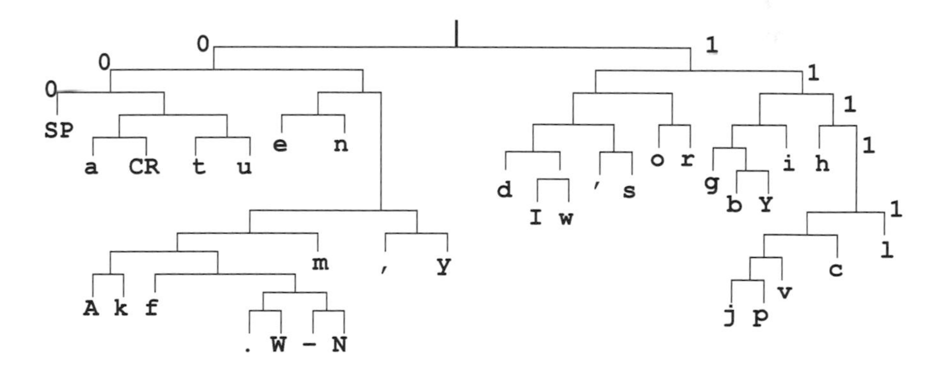

FIGURE 9.7 An Example Huffman Code Tree

right branch, we see the most frequent character (SPACE) can be represented with just 3 bits, 000.

Following the code tree, we see the code for the first three characters (You) is 110011101000111.

The resulting Huffman code for the entire poem is 2457 bits, as compared to the ASCII file, which is $545 \times 8 = 4360$ bits.

Of course, we also need to include the required Huffman tree as part of the deflated message so it can be inflated later. This costs no more than

$((number\ of\ symbols) \times (sz + 2) - 1)$ bits, where sz is usually 8 in ASCII code. For our example, the cost is 339 bits.

Applying the GZIP tool on this poem, we get a file 291 bytes long, which is still better than using Huffman code alone (**Demo:** `TryGZIP`).

> **CT: Customize for Efficiency**
>
> *One size does not fit all well. Custom solutions can examine and take advantage of properties in individual cases and become more efficient.*

Representing a message as a stream of characters with ASCII/UNICODE encoding fits all. Yet, data compression schemes can examine individual messages and deflate each one using repeated patterns and character frequencies unique in the message.

Modern Web servers can automatically deflate specific types of documents, such as HTML, CSS, and JavaScript, just before sending, and Web browsers can automatically inflate them when received. This, working together with HTTP caching (Section 6.10.1), can significantly speed up loading of Web pages and reduce unnecessary Internet traffic.

9.9 Data Structures

As you may expect, a computer program must usually manipulate data to achieve its goals. An important task in any program is to organize and manipulate data correctly and efficiently.

The term *data structures* in computer science refers to well-known techniques to organize data often used in programming. These structures provide systematic ways to store and manipulate in-memory data.

For example, a *tree structure*, or tree for short, is useful for organizing data in a hierarchy. We have seen the file tree (Section 4.8), the solution tree (Figure 8.11), and, of course, the Huffman code tree in the previous section (Section 9.8.2).

Here is a list of common data structures:

- Array—A sequence of equal-size memory cells in consecutive memory locations. Each cell can be reached directly by an integer index, starting from zero. An array stores data items of the same type, an array of integers, characters, or names, for example (Section 3.5.2).

- Linked list—A sequence of data items where each item has a pointer to the next item. The last item has points to a special end symbol. A list makes it easy to splice in or cut out an item. Each item on a bidirectional list also has a pointer to the previous item.

- Queue—A specialized list or array where items are processed in a first-in-first-out order (FIFO).

- Stack—A specialized list or array where items are processed in a first-in-last-out order (FILO), like a stack of plates in the cafeteria (Figure 9.8).

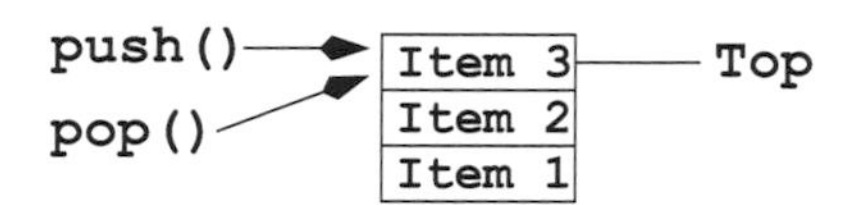

FIGURE 9.8 A Stack

- Tree—A hierarchical structure with a root node pointing to child nodes that may point to their own child nodes. A node with no children is called a leaf node. Each node, except the root, has a unique parent node. The Huffman code tree is a binary tree (up to two child nodes).

- Buffer—A holding place for data items, where a producer program can deposit data and a consumer program can remove data.

- Table—A collection of *key-value* pairs, not unlike a dictionary. A table makes it easy to look up values by keys. A *hash table* is a particularly efficient way to store and retrieve items by computing *hash codes* for keys allowing direct access to items stored in the table without searching through the table.

- Graph—A structure with nodes connected by edges. Consider a graph as a generalized tree where any node can point to any other node, not just to child nodes. This structure is useful to represent connectivity in networking, transportation, and so on.

As you can see, each data structure provides two kinds of programming support: data organization and data manipulation. For example, a stack organizes data items in a linear sequence and then it provides *push* (onto the top), *pop* (off the top) and *is-empty* (test) procedures for using a stack. To protect the integrity of a stack, these are the only allowed operations.

CT: Synthesize and Simplify

Synthesize lower-level details and operations into understandable higher-level objects and actions. They can simplify complicated tasks.

Such data structures help hide data and operational details and allow programmers to think about, for example, pushing and popping a stack instead of bit patterns, memory addresses, and other such details. This is a form of abstraction (CT: Abstract Away, Section 1.2) known in computer science as *data abstraction*.

9.10 What Is a Database?

Large, complicated, and dynamically changing data, such as student records, store inventories, and airline reservations, can not be adequately handled in static files, such as spreadsheets. They require database systems.

A database is a collection of data efficiently organized in digital form to support specific applications. A database is normally intended to be shared and updated by multiple users.

A *Database Management System* (DBMS) is a software system that operates databases, providing storage, access, security, backup, and other facilities. A DBMS usually makes databases available to many users at the same time and handles user login, privileges (abilities to perform certain operations), and local or network access.

A *Relational Database* is one that uses *relations* or tables to organize data. Even though there are other ways to organize data, relational databases are by far the most popular and widely used in practice. Well-known relational DBMS (RDBMS) include IBM DB2, Oracle, MySQL, SQLite, Derby, and Microsoft Access.

9.10.1 Relational Databases

A relational database uses multiple tables, called *relations* in database theory, to efficiently organize data. A *relation* is a set of related attributes and their possible values. Figure 9.9 shows a typical database table.
Each table is defined by a *schema* that specifies

1. The names of the *attributes* (column headings)

2. The type of value allowed for each attribute

Last	First	Dept	Email
Wang	Paul	CS	pwang@kent.edu

FIGURE 9.9 A Simple Database Table

For example, the type of an attribute can be character string, date, integer, or decimal, and so on.

Each row in a table is a set of attribute values. A row is also called a *record* or a *tuple*, and no two records can be exactly the same in a table (the *no duplicate row* rule). The collection of all records in a table is called a *table instance* or *relation instance*. Immediately after a table is defined by its schema, it has no records. So the table instance is the empty set. As records are inserted into the table, the table instance grows. As records are removed from the table, the table instance shrinks. As data in table records are changed, the table instance changes. When dealing with database tables, it is important to keep in mind the difference between the table schema and the table instances.

Each database in an RDBMS is identified by a name. A relational database usually consists of multiple tables organized to efficiently represent and interrelate the data. The RDBMS also stores, in its own management database, information about which users can access what databases and what database operations are allowed. Usually, a user must first login to the RDBMS from designated hosts before access can be made. For example, if access is restricted to the *localhost*, then only access made by a program, running on the same host computer as the RDBMS, is allowed.

9.10.2 SQL: Structured Query Language

Of course, each RDBMS needs to provide a way for application programs to create, access, and manipulate databases, tables, records, and other database-related items. The *Structured Query Language* (SQL[2]) standardizes a programming language for this need. SQL, an ISO (International Organization for Standardization) and ANSI (American National Standards Institute) standard, is a *declarative language* that uses sentences and clauses to form *queries* that specify database actions. SQL consists of a *Data Definition Language* (DDL) to specify schemas, and a *Data Manipulation Language* (DML) for adding, removing, updating, retrieving, and manipulating data in databases.

Major RDBMSs are all SQL compliant to a great degree. Thus, programs

[2]SQL is pronounced by saying each of the three letters.

coded in SQL can easily be made to work on different systems. However, there are still differences among different database systems. In this book, examples show SQL code for the MySQL system. The code can easily be repurposed for other RDBMSs.

A *database query*, or simply a query, is a command written in SQL that instructs the RDBMS to perform a desired task on a database. A data retrieval query usually returns a *resultset*, which is a table of records. A data update query does not return a resultset.

Today, the Web and database systems are nearly inseparable (Section 6.8.2). The Web supercharges the power of databases by providing them with on-Web frontends (Figure 9.10). Any untrained user can fill forms in webpages to retrieve and update information in databases. At the same time, databases support various on-Web operations, such as user accounts, making them more functional and useful. As a result, you have online class registration, shopping, banking, ticketing, investing, and many other functions most of us can't do without in our daily lives.

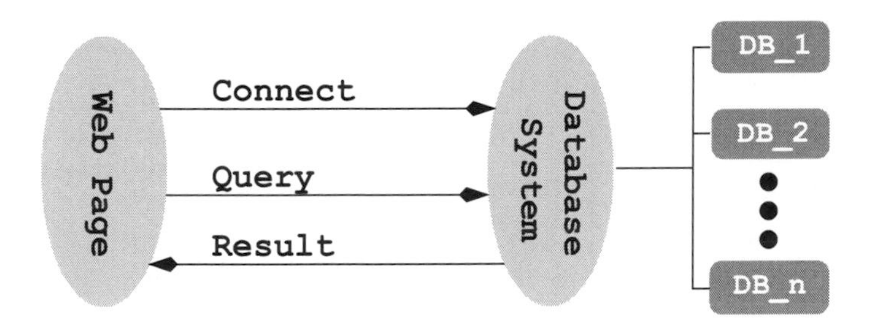

FIGURE 9.10 Web Access to Databases

> **CT: Combine Web and Database**
>
> *Combine the Web and database systems. Form a powerful union much greater than its parts.*

Modern technologies, such as space probes and telescopes, earth resource and observation satellites for example, generate tremendous amounts of data. It takes enormous computing power to digest and understand such data in order to discover knowledge from them. Database systems also play an important part in such efforts.

9.10.3 Big Data

Advances in digital and information technologies, especially in the sensing and gathering of data in many different areas, result in collections of data that are extremely large. It was estimated that, as of 2012, 2.5 exabytes (10^{18} bytes) of data were created every day! And the data collection rate has been increasing rapidly as well.

Basically, we have the ability to collect data on everyone, everything, everywhere. And we do it in many different fields including census, demographics, medicare, weather prediction, crime fighting, science, business, and finance. Very large databases are needed to store big data. Working with databases of such sizes requires parallel programs executing on hundreds or thousands of computers.

CT: DATA TO INSIGHT

Big data is big resource. Analyzing big data can reveal hidden relations and correlations, which can be invaluable in decision and policy making.

The breakthrough thinking is "We are not limited to sampling data and statistics. We have the abilities to collect, examine, and analyze, all the data."

9.11 Protecting Personal Data

The widespread creation, collection, transmission, and access of data online have made the information age what it is. The benefit it brings is immense and indispensable.

However, we should also be aware of the drawbacks. Particularly, we need to prevent *personal data* from falling into the wrong hands or being misused. Examples of personal data are full name, date and place of birth, ID numbers, address, phone number, login information, screen name, account numbers, financial and credit information, medical records, health information, and so on.

CT: GUARD PERSONAL DATA

Individuals and organizations must do their best to keep personal data confidential.

Generally, personal data is any information that can be identified with, is specific to, or can be connected to a living individual. Usually, personal data must be kept confidential by any individual or company that collects or retains the information. Many countries have regulations that define personal data and the responsibilities collectors and keepers have to protect such information.

Exercises

9.1. What is the resolution of an HD monitor?

9.2. If each pixel requires 24 bits, what is the storage size of a raw HD image?

9.3. Explain *additive color* and *subtractive color* models.

9.4. In printing, why do we need black ink when combining CMY inks can produce black?

9.5. In 24-bit RGB notation, what color is (255,0,255)? (100, 100, 100)? (500,400,600)?

9.6. Why do we want to reduce the resolution when taking pictures with our digital cameras or smartphones?

9.7. What is the difference between vector and raster graphics?

9.8. What is sampling? Quantization? How are they related to the final data size?

9.9. What is a codec?

9.10. In an application program, what is the usual meaning of these operations: **save**, **save as**, **export**, **import**.

9.11. What is document markup, and why is it useful?

9.12. What is data compression? Lossy and lossless compression?

9.13. Compress the SVG code for the `webtong.com` logo (Figure 9.2), in Section 9.2.3), using both GZIP and ZIP. Report how small the compressed files are.

9.14. What are the two main techniques to deflate a message?

9.15. Refer to Figure 9.7 and deduce the code word for each of these characters: `b`, `Y`, `y`, and `f`.

9.16. Continue from the previous exercise and write down the code words for all 34 characters.

9.17. Look at the result in the previous exercise. Do you see any code word that happens to be a prefix of some other code word? If not, please explain why this is always the case for Huffman code.

9.18. Consider the following bit encoding of a Huffman tree.

Algorithm `treeEncode`:
Input: Root node `nd` of tree to be encoded
Output: Bit string to represent the tree

(a) If `nd` is leaf node, then output a 1-bit followed by the 8-bit character/byte and return.

(b) Output a 0-bit; call `treeEncode(left child of nd)`; call `treeEncode(right child of nd)`.

What is the bit length of this bit string? How would you decode this bit encoding back into a binary tree with characters at the leaves?

9.19. What is a database, a table, and a relation?

9.20. What is SQL, and who needs it?

9.21. **Computize**: Find out about the current state of digitizing the human smell, taste, and touch.

9.22. **Computize**: TV screens are going to 4K. What resolution is that? What implications are there for streaming shows in 4K?

9.23. **Computize**: Look into 3D printing, how it works, and how it relates to representing digital images.

9.24. **Computize**: If bit patterns are the only way to store information, how do you think executable programs are stored? How are they distinguished from non-executable files? And who cares?

9.25. **Computize**: Invent a personnel record markup. Give an example record in your markup.

9.26. **Group discussion topic**: *QR codes and their uses.*

9.27. **Group discussion topic**: *My favorite search engine.*

9.28. **Group discussion topic**: *The way Google Translate works and improves through time.*

Chapter 10

Get That App

Programs make computers work. Without programming, a computing device might as well be a brick.

The master program is the operating system (OS; Chapter 4), which controls everything, including hardware, files, networking, users, and other programs that run on a computer.

An application program, or app, is a self-contained software program that, by interacting with the user, performs a specific task. Some applications are intuitive to use, but most also require some learning and skills. Together, they make computers useful in countless ways and life easy and enjoyable. Other programs do not interact with users directly but support the inner workings of a computer, such as networking and peripheral device control. Understanding how to deal with programs effectively and apply them efficiently will be important. Getting a program involves the physical, downloading and installing, and the intellectual, learning and understanding.

Large computer programs are among the most complicated tools humans have ever created. An understanding of programming technologies and how programs are developed is increasingly important in the digital age.

In the beginning of this book (Chapter 1), we learned that one of the most significant aspects of the digital computer is its ability to load and execute different programs. That ability makes it a universal computing machine, as defined by the Turing Machine model (Figure 1.2). Modern computers store programs in RAM (Random Access Memory) for fast execution.

Basically, a program is a complete set of instructions, written in a certain *programming language*, to perform well-defined tasks. Install the program onto a computer, run the program, then the computer becomes a machine that functions (operates) as specified by the program. On our smartphones, we take all this for granted—download, install, and run an app, and instantly the phone becomes a calculator, a GPS navigator, So convenient and so universal, don't you just love it? It is magical!

Furthermore, special-purpose computers make everything *smart*: automobiles, airplanes, household appliances, medical scanners, and robots. Often, they also have reprogrammable firmware and updatable software. There is a lot behind the install-and-run magic on computers, and much of it is due to advances in computing technologies and programming methodologies.

An appreciation of programs, programming, how to effectively use programs, as well as ways programs are developed is an essential part of understanding computing.

10.1 Key Programs

On a computer, OS is obviously the most important program. Without it, your computer simply does not work at all. Other indispensable programs that make a computer useful and powerful include:

- Web browsers—For surfing the Web (Chrome, Firefox, IE, Safari, for example)
- Email clients—For sending and receiving email (Outlook, Gmail, Thunderbird, for example)
- File managers—To search and manage all your files stored in your file system (Windows File Explorer, Nautilus, XFile, for example)
- GUI elements—To support graphical user interface needs (Section 4.4), including window drawing, desktop (home screen) display, event handling, application window management, and informational widgets (clock, calendar, weather, for example).
- Text editors—For creating, editing, and displaying plaintext files (Wordpad, Vim, Emacs, TextEdit, for example)
- Office productivity tools—To make office/school work easy and efficient (MS Office, Libre Office, NeoOffice, Adobe Acrobat, TeX/LaTeX, for example)

Additional *applications* or *apps* are available for many different purposes. Widely used apps include media players (pictures, music, video), such as iTunes, and YouTube; games, such as Angry Birds; weather report/forecast; and maps for directions and navigation, such as Google Maps.

An app may be designed to work on desktops, laptops, tablets, or smartphones. Most will work offline, others require an Internet connection to function fully, and still others, such as Web-based email apps, are completely online. Here are some additional apps you may find useful:

CT: Remind Yourself

We are all forgetful to some degree. But the computer can extend our memory. Use a calendar reminder program/service and you'll never forget anything important again.

Applications such as Windows Calendar, Windows Outlook, Mac iCal, Google Calendar, as well as calendar, and reminder on Linux. These apps allow you get email or smartphone notification reminders. And you usually can set the lead time for receiving your reminders. It is hard to think of a reason not to use a calendar reminder.

Here are some additional apps you may find useful.

- Secure remote login—Applications, such as SSH, WinSSH, OpenSSH, that allow you to connect to another computer via the Internet, login, and use that remote computer from where you are, usually via a CLI (command-line interface; Section 4.7).
- Secure FTP—You can securely upload and download files to/from another computer with programs such as SFTP, FileZella, and FreeFTP. Usually a secure remote login app also supports FTP.
- Making calls—Skype, Google Hangout, and similar apps allow you to make free (phone) calls to others via the Internet, optionally with video.
- Image processing—Gimp (free) and Photoshop (commercial) are two of the best photo editing and image processing tools.
- Video and audio editing—Movie Maker (Windows), iMovie (Apple), Premiere Pro (Adobe), OpenShot (Linux) are examples.

CT: Install That App

Want to do XYZ? You can often find one or more apps for it to install and run on your system.

If you need to perform some task and you don't have an app on your computer for it, chances are you can find that application and install it in a matter of minutes. This is especially true on smartphones. For example, small business owners can install an app and take credit card payments on their smartphones. Such apps include Square Register™, PayPal Here™, Payleven™, and PayAnywhere™.

Be careful installing software. Make sure the program is from a reputable source. Innocent-looking apps may contain malware (Section 7.10) that can seriously damage your computer.

10.2 Knowing Your Apps

A set of application programs comes with your computer system. There may be many you don't need. Consider uninstalling them. Modern apps will automatically update themselves. You can also check and manually update them, usually via the program's `help` menu. It is advisable to download and install OS updates regularly, especially to fix security issues.

Usually an application can be started (launched) in several ways:

- Clicking an icon on the taskbar starts/resumes an app.
- Double clicking an icon on the desktop launches an app.
- Clicking on a name in a list of available programs starts an app.
- Clicking or double clicking a data file opens the *default app* associated with that type of file (by the OS).
- Right clicking, or control-clicking on Mac OS X, a data file and selecting an app opens that app.
- Clicking a data file link in a webpage opens an app associated with the particular file type (by the Web browser).
- Opening an email attachment launches an app associated with that type of attached file (by the email client).
- Giving a Shell command (Section 4.7) starts/resumes apps in many ways.
- Clicking a URL in an email or other document launches the default Web browser configured in the OS.

You may have multiple applications running at once. Resume (return to) any running application by bring its window into focus (Section 4.4) or by clicking its representation on the taskbar. You may also maximize, unmaximize, resize, and minimize application windows. And you can group windows in distinct *desktop workspaces* to efficiently use the available display space.

Even though you can hold on to multiple app windows and resume any when you desire, it is good to close an app when you are finished with it. This can be done by closing the app window or selecting the `quit` option on its `file` menu. You can also use the operating system task manager to "force quit" any nonresponsive app.

CT: LEARN THAT APP

Apps are tools. Skillful use can pay big dividends. Practice makes perfect.

Each app provides its own distinct interface for users to effectively control its functions. Apps strive to provide a clear, intuitive, easy to use, and visually pleasing user interface. Still, keep in mind that each app speaks a language of its own and, generally, an operation in one app won't work in another app. In fact, it pays to become skillful in using apps you often need. Start by learning well your favorite Web browser.

An app also depends heavily on its operating environment, which is usually provided solely by the operating system (Figure 10.1).

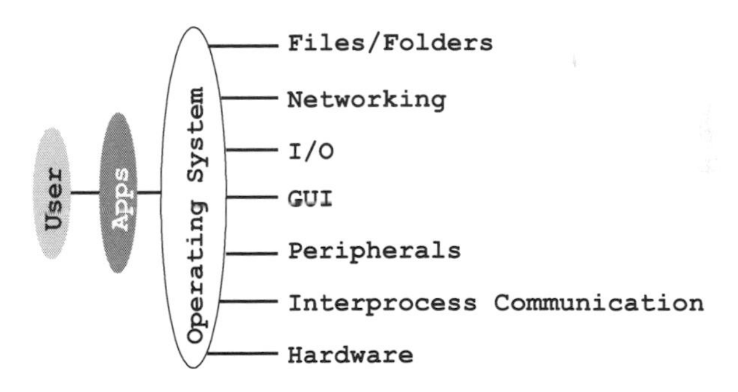

FIGURE 10.1 Using A Computer

The OS provides file access, networking, I/O (input, output) operations, GUI (graphical user interface) support, interprocess communication (Section 10.4), and *device drivers* for control of peripheral devices (USB devices, for example). When obtaining a new app, we need to get the one designated for our operating system, or it won't work at all.

CT: NO APP, NO WAY

You don't use a computer. You use a computer only through some program.

10.3 Program Configuration and Customization

An important advantage of software over hardware is flexibility. Software can be easily configured, fine-tuned, and customized by users to work in different environments and to satisfy particular preferences.

To begin using a new computer, you immediately set the language, locale, time zone, host name, work group, and other parameters to customize the operating system. You can also pick a preferred cursor style, font size, resolution, and gamma setting for your display. Register default applications, such as Web browser and email client, with your OS. And you may add support for additional languages together with input methods.

Similarly, each application program allows you to set preferences, select options, and customize it for effectiveness and convenience. Here are a few examples:

- Setting homepage and other preferences for your Web browser; picking your preferred search engine
- Adding your personal signature card, contact list, and perhaps also message encryption keys to your email client; setting plaintext or HTML as your default email sending format
- Setting the font size, color, as well as background and size for your Shell window

> **CT: Configure and Enjoy**
>
> *It pays to configure and customize your system and apps.*

It is important to configure your app to take full advantage of its flexibility and power. It takes time and effort to learn about the available options and their effects. Take the camera app on your smartphone for example. Have you explored its varied options? Or do you just point and shoot?

10.4 Process Cooperation

We already know that a program in execution is called a process (Section 4.9). Many processes are usually running concurrently, or actually in parallel when a computer has multiple CPUs (cores). A process interacts with users through a GUI or CLI. A process can also interact with other processes through *inter-process communication.*

Basically, running programs can communicate with one another by sending data back and forth. For example, a program can invoke another program (a *child process*) and send input to it and obtain output from it. Interprocess communication allows us to combine the abilities of different programs to solve problems. The desktop clipboard (Section 4.5.1) is a simple example.

A *pipe* is a memory area where one process stores data and another process retrieves data. Figure 10.2 shows the way two processes can perform interprocess communication. By combining and reusing existing programs, we avoid reinventing the wheel and save both time and effort.

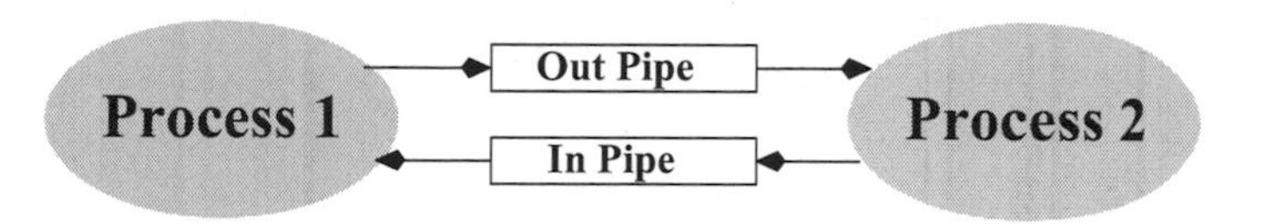

FIGURE 10.2 Interprocess Communication

> **CT:** COORDINATE OR ELSE
>
> *Cooperation among concurrent processes depends on precise coordination to avoid problems and work smoothly.*

When cooperating, the process sending data is known as a *producer*, and the process receiving data is known as a *consumer*. To maintain smooth flow of data through the shared area (the pipe for example), the producer and consumer processes must coordinate and synchronize their actions. They must also wait for data to become available or for the shared area to be vacated before proceeding.

Unix/Linux and Microsoft Shells allow users to set up *pipelines*, sending output of one process as input to the next process. Here is an example pipeline:

(*listing file names in a certain folder*) | (*sorting input into order*),

where the VERTICAL BAR character (|) is used to indicate a pipe connection. Such CLI flexibility and power are not available to users under GUI.

Processes can also send *signals* or *interrupts* to one another. A signal or interrupt can get the attention of a running process outside of its normal control flow. The situation can be likened to calling someone on the phone. The person must stop what he or she is doing and answer the call. Similarly, a process must stop what it's doing and react to an incoming signal or interrupt. The receiving process will normally execute a predetermined action in response to a particular signal. It then can either resume or terminate execution.

10.5 Machine Language Programs

All these programs are wonderful. But how are they developed? In the information age, developing programs is just as important as producing goods. Let's turn our attention to how computer programs are written.

A program needs to be written in a language that is understood by a particular computer. The native language of a computer, its *machine language*, is defined by the *instruction set* of its CPU (Section 2.1). In the early days of computers, before better programming tools were introduced, programs could only be written in machine language.

A machine language program consists of a set of instructions each of which usually involves:

- An *opcode*—An opcode is a bit pattern specifying a particular elementary operation for the CPU to perform, such as addition, subtraction, xor, or go to another instruction.
- One or more *operands*—An operand is a CPU register or memory address where data or instruction can be found or stored. It can also be a constant value.

Opcodes instruct the CPU to set register values, retrieve/store data or instruction from RAM, perform arithmetic and logical operations, and go to a different memory location for the next instruction (branch). Figure 10.3 shows the CPU in action.

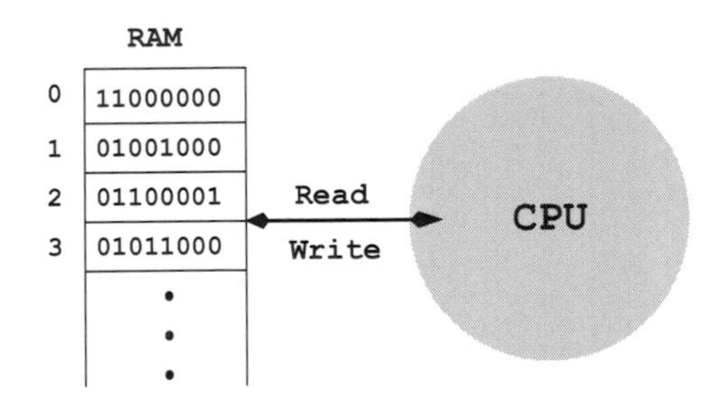

FIGURE 10.3 CPU Instruction Execution

Consisting of nothing but `0`s and `1`s, machine language programs are very difficult to write directly, as you can imagine. However, because the opcodes correspond directly to the CPU's instruction set, machine language programs are easy on the computer. Of course, a machine language program will not run on any other computer with a different CPU.

As a concrete example, Table 10.1 illustrates the machine code for the assignment `A = A + B`, where the opcodes are, in turn, for `LOAD` (load value at given address into cpu register), `ADD` (adds value at given address to register), and `STORE` (stores value in register back to given memory location).

Today, there is rarely a need for programmers to write programs in machine language.

TABLE 10.1 Sample Machine Language Program Fragment

Opcode	Memory Address	Assembly Code
110000000	00100000 00000000	LOAD A
101100000	00010000 00000000	ADD B
100100000	00100000 00000000	STORE A

10.6 Assembly Language Programs

An assembly language makes programs in a particular machine language slightly easier to write. An assembly language defines a symbol, known as a *mnemonic*, such as `ADD`, `LOAD`, and `STORE`, for each opcode in the machine instruction set. Furthermore, it allows symbolic labels, such as `A` and `B`, and even expressions to be used as operands. In Table 10.1, simple assembly language equivalents for the machine code are also shown.

Modern assembly languages reflect the advances of processor architectures. For example, the popular Intel x64 (x86_64) processors have sixteen 64-bit registers, virtual memory, as well as context switching support (Section 4.9). Multicore processors can execute in parallel several independent instruction streams (one per core) and therefore provide hardware support for multiprocessing (Section 4.9).

Table 10.2 shows the first four 64-bit, general-purpose registers with various standard names used to reference them in x64 assembly language.

TABLE 10.2 First Four x64 Registers

64-bit Register	Lower 32 Bits	Lower 16 Bits	Lower 8 Bits
rax	eax	ax	al
rbx	ebx	bx	bl
rcx	ecx	cx	cl
rdx	edx	dx	dl

An assembly language program is translated into machine language by a utility program known as an *assembler* that produces machine code for a particular computer system (Figure 10.4).

Generally speaking, an assembly language and its assembler are dependent on the target processor as well as on the target operating system. For example, it makes sense to talk about x64 assembly under Windows or Linux.

Today, writing programs in assembly is rare. Only device drivers, embedded systems, and real-time applications may require hand assembly language

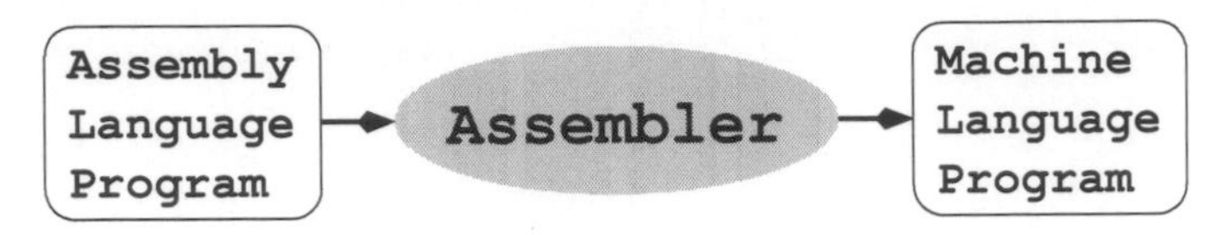

FIGURE 10.4 Assembler

coding in some critical places. Mostly, programs are written in high-level languages and *compiled* into machine code. Usually, an assembler would function as part of the compiling process.

10.7 High-Level Programs

Over the years, programming technologies have evolved and matured in many ways. Today, high-level languages make creating programs much faster. They hide low-level details, those relating to hardware and operating systems, and present a paradigm, an abstraction (CT: ABSTRACT AWAY, Section 1.2), for programmers to easily and efficiently think, reason, and create code. We first mentioned high-level languages (Figure 1.8) and programming in Section 1.5. A high-level program can be executed on a computer in two possible ways:

- Compilation—The high-level program is translated, by a *compiler* program, into machine language that can be executed directly on a given computer.
- Interpretation—The instructions in a high-level program can be read and understood by an *interpreter* program, which will in turn carry out the needed actions.

This means compilers or interpreters must be developed for each particular computer platform that will run programs in a particular high-level language.

The first widely accepted high-level language is *Fortran* (Formula Translating System), developed originally by John W. Backus at IBM in the 1950s. Fortran introduced constructs for procedures, functions, and loops that are user (programmer) oriented. Such high-level constructs make Fortran much easier to use by programmers than assembly languages (CT: DEVELOP FOR USERS, Section 6.11). Fortran is particularly good for computing with numbers (integral and floating-point) and therefore great for engineering and scientific computations. It has been so successful that it has been in continuous use ever since. The language has been evolving along the way (FORTRAN 77, Fortran 90, Fortran 95, Fortran 2003, and Fortran 2008) and is easily available on most computers.

Fortran introduced the first compiler that *optimizes*, improves the efficiency of, the generated code. The Fortran compiler represents a significant breakthrough in programming technology.

C, arguably one of the most successful programming languages ever, was developed by Dennis Ritchie at Bell Labs in the early 1970s under the Unix operating system. For example, the recursive gcd algorithm we described in Section 8.3 can be written in C as

```
int gcd(int a, int b)
{   if ( b == 0 ) return(a);
    else return ( gcd(b, a % b) );
}
```

As you can see, it is not hard to read or understand. Check out the compiled assembly code, **Demo:** GCDCode at the CT site, with 37 lines that are much harder to read.

Later, mainly by adding the *class* construct for *object-oriented programming* to C, Bjarne Stroustrup created the C++ language. Today, C++ is considered a superset of C and ISO[1] standard C/C++ is used widely for procedural programming, as well as object-oriented programming (OOP; Section 10.10).

There are many other high-level languages. Some will be mentioned later in this chapter.

Generally speaking, high-level languages can also help programs achieve many desirable qualities, including:

- Readability—Easy to understand and follow
- Modular structure—Organized into small, self-contained modules with well-defined application interfaces
- Portability—Easily made to work on different systems or platforms
- Flexibility—Simple to reconfigure to address changing application contexts and requirements
- Modifiability and maintainability—Reducing the effort for making corrections, changes, modifications, and updates
- Reusability—Readily applied in similar situations as a program component, avoiding reinventing the wheel

High-level programming languages are also characterized by various properties that distinguish them from others.

- General-purpose or special purpose—A general purpose language is for writing almost any kind of program, while a special purpose language is for use in specific domains or areas. For example, C/C++, Java and Fortran are general purpose, while PHP (for Web server), JavaScript (for Web browser), and SQL (for relational databases) are special purpose.

[1]International Organization for Standardization

- *Imperative* or *declarative programming*—An imperative program gives a sequence of steps, a procedure, to be performed (C and Fortran, for example). A declarative program specifies the logic and desired results without detailing a procedure to follow. For example, in SQL, queries such as, "*Select Freshman students whose age is between 18 and 20 inclusive*," are used.
- *Sequential* or *parallel*—A language is sequential (most are) if it allows a single control flow or execution thread. It is parallel if it supports more than one control flow or multiple simultaneous/concurrent execution threads. Java is a language with such features.
- *Procedural* or *object-oriented*—A language with or without support for defining and instantiating run-time objects (See Section 10.10).

10.8 Compilers

A *compiler* is a program that translates a high-level language program into machine language executable on a target computer (Figure 10.5).

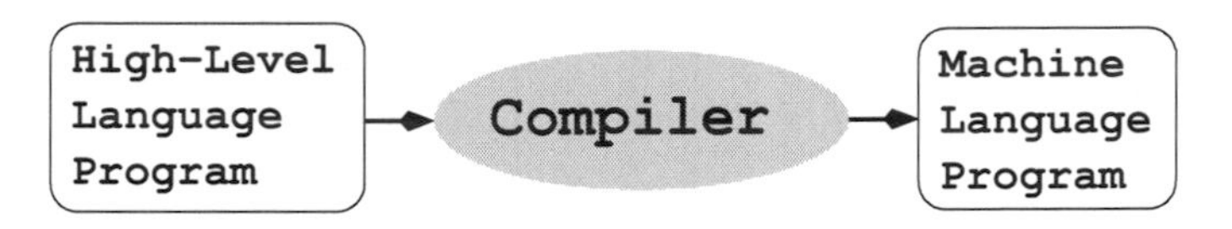

FIGURE 10.5 Compiler

In addition to code translation, a compiler also takes care of arranging suitable *run-time support* for the program by providing I/O, file access, and other interfaces to the operating system. Therefore, a compiler is not only computer hardware specific, but also operating system specific.

Let us look at GCC (the Gnu C Compiler) as an example. GCC breaks the entire compilation process into five phases (Figure 10.6).

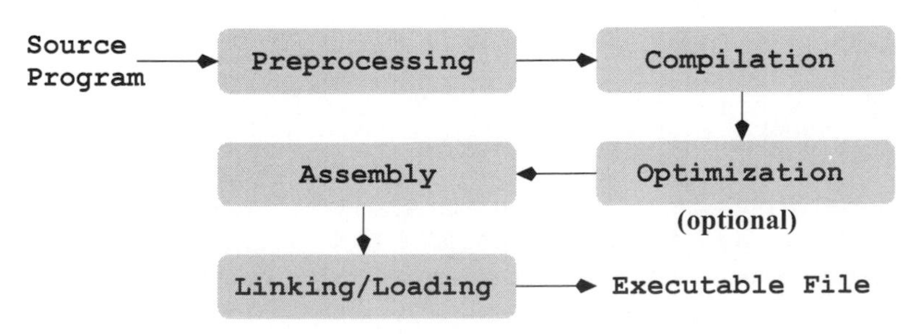

FIGURE 10.6 Compilation Phases

1. *Preprocessing*—The first phase is performed by the *C preprocessor* program that handles constant definition, macro expansion, file inclusion, conditionals, and other preprocessor directives in the program source.

2. *Compilation*—Taking the output of the previous phase as input, the compilation phase performs syntax checking, parsing, and assembly code generation.

3. *Optimization*—This optional phase specializes the code to the computer's hardware architecture and improves the efficiency of the generated code for speed and compactness.

4. *Assembly*—The assembler program takes assembly code input and creates *object files* containing binary code and relocation information to be used by the linker/loader.

5. *Linking*—The *linker/loader* program combines all object files and links in necessary library and run-time support programs to produce an executable program.

As you can see, a C compiler is a complicated program. In what language is it written? Well, it is written in C! Before you have a C compiler you cannot write any program in C. Once you have a working C compiler, you can write any program in C, including a C compiler.

CT: BOOTSTRAPPING

> *Take a small step forward, build on that progress to make further advancements. The feedback loop can lead to very advanced systems quickly.*

In computing we call this *bootstrapping*. The evolution of programming environments is an excellent example of bootstrapping. Starting with a simple text editor and basic assembler, we can build better text editors and assemblers, then a simple compiler for a higher-level language (FORTRAN) that allows us to write more complex code for compilers to support even higher-level languages, and so on. Today we have comprehensive IDEs (Integrated Development Environments) that combine many useful tools for program development in a particular language with easy-to-use GUIs. IDEs can greatly facilitate program coding, version control, testing, debugging, documentation, and project management.

10.9 Software Development

In the digital age, much effort goes into the creation of software products. Software development involves conception, research, requirement analysis, program design, program implementation, testing, debugging, documentation, maintenance, feedback collection, updates, and new releases.

As you would expect, a variety of software tools have been created just to support these software development activities.

Object-oriented design and programming are well-established technologies for the design and implementation of high-quality software.

10.10 Object-Oriented Programming

Achieving desirable program qualities (Section 10.7) is ever more important as computer programs become larger and more complicated. To this end, we need to better structure and organize large programs.

> **CT: COMPARTMENTALIZE**
>
> *Separate a complicated system into isolated, self-contained, and replaceable parts.*

Modular programming is a software design technique that promotes the organization of the entire program into task-specific, self-contained, independent *modules* that can be easily plugged or unplugged from the larger program. All code and data related to the particular task are gathered into one module and mostly hidden from external view. Each module *exposes* an API (Application Programming Interface) that specifies how other programs that apply the module may interact with it.

Object-Oriented Programming (OOP) is a widely used programming paradigm that supports not only modularity but also many other desirable qualities for good programs.

The central idea of OOP is to build programs using software *objects*. An object can be considered as a self-contained computing entity with its own data and programming. On modern computers, windows, menus, and file folders, for example, are usually represented by software objects. But objects can be applied to many kinds of programs. An object can be an airline reservation record, a bank account, or even an automobile engine. An engine object would include data (called *fields*) describing its physical attributes and programming (called *methods*) governing how it works internally and how it interacts with other related parts (also objects) in an automobile.

A payroll system would have employee records, time cards, overtime, sick leave, taxes, deductions, etc., as objects. An air traffic control system would have runways, airliners, and passenger gates as objects. Thus, in OOP, the software objects correspond closely to real objects involved in the application area. This correspondence makes the computer program easy to understand and manipulate. In contrast, traditional programming deals with bytes, characters, variables, arrays, indices, and other programming artifacts that are difficult to relate to the problem at hand. Also, traditional programming focuses mainly on the step-by-step procedures, to achieve the desired tasks. For this reason, traditional programming is also known as *procedure-oriented* programming.

10.10.1 OOP Advantages

OOP offers the following main advantages:

- Simplicity: Because software objects model real objects in the application domain, the complexity of the program is reduced and the program structure becomes clear and simple.
- Modularity: Each object forms a separate entity whose internal workings are decoupled from other parts of the system.
- Modifiability: It is easy to make minor changes in the data representation or the procedures used in an OO (Object Oriented) program. Changes within an object do not affect any other part of the program, provided that the *external behavior* of the object is preserved.
- Extensibility: Adding new features or responding to changing operating environments can be a matter of introducing a few new objects and modifying some existing ones.
- Flexibility: An OO program can be very flexible in accommodating different situations because the interaction patterns among the objects can be changed without modifying the objects.
- Maintainability: Objects can be maintained separately, which makes locating and fixing problems and adding "bells and whistles" easy.
- Reusability: Objects can be reused in different programs. A table-building object, for instance, can be used in any program that requires a table of some sort. Thus, programs can be built from prefabricated and pretested components in a fraction of the time required to build new programs from scratch.

C++ and Java are well known and widely used high-level languages that support OOP. C++ is compiler based. Java is interpreter based. The Java Virtual Machine (JVM) is a program that interprets Java byte code. A Java program can run on any platform that has a JVM.

10.10.2 OOP Concepts

The breakthrough concept of OOP technology is the attachment of procedure code to data items. This concept changes the traditional segregation between data and procedures. The wrapping together of procedures and data structure is called *encapsulation*, and the result is a software object.

In an OOP language, such as Java, all procedures are encapsulated, and they are called *methods*. For example, a window object (Figure 10.7) in a

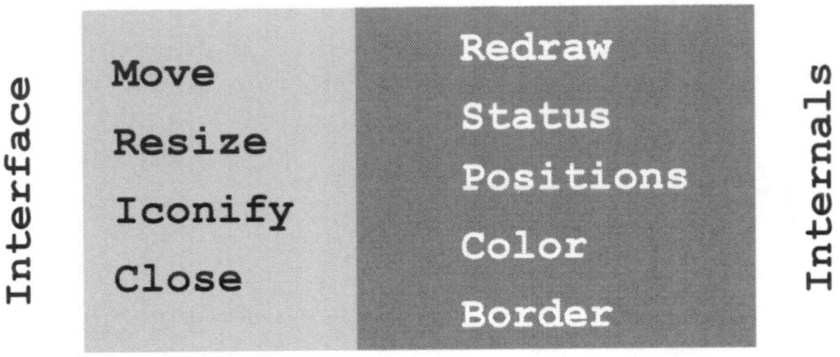

FIGURE 10.7 A Window Object

graphical user interface system contains the window's physical dimensions, location on the screen, foreground and background colors, border styles, and other relevant data. Encapsulated with these data are methods to move and resize the window itself, to change its colors, to display text, to shrink into an icon, and so on.

Other parts of the user interface program simply call upon a window object to perform these tasks by invoking the window object's exposed interface methods. It is the job of a window object to perform appropriate actions and to keep its internal data updated. The exact manner in which these tasks are achieved, and the structures of the internal data are of no concern to programs outside the object. The object's interface completely defines how to use that object. This interface is the *application programming interface* (API) of the object.

> **CT: Expose Only the Interface**
>
> *Hide internal workings from clients and expose only ways to interface to your service.*

The hiding of internal details is a form of abstraction (CT: Abstract Away, Section 1.2). The API is the exposed interface to clients. Compartmentalization often results in a better organized system that is easier to maintain, manage, test, and improve. Problems can be traced to individual parts and fixed easily. OOP allows us to do exactly this. See **Demo:** `SampleJavaAPI` at the CT site for an example.

The separation of API from *internal workings* is not difficult to understand. In fact, it is common practice in our daily lives. Consider a bank teller, for example. Customers go to any bank and talk to any teller using the same set of messages: account number, deposit, withdrawal, balance, and so on. The way each bank or teller actually keeps records or performs tasks internally is of no concern to a customer. These tried-and-true principles simplify business at all levels and can bring the same benefit to organizing programs.

As an OO program executes, objects are created, messages sent, and objects destroyed. These are the only allowable operations on objects. The internal (*private*) data or methods in an object are off limits to the *public*. The decoupling of the private mechanisms in objects from routines outside the objects significantly reduces the complexity of a program.

It is often the case that more than one object of the same type is needed. For example, multiple windows often appear on workstation screens. Normally, objects of a given type are *instances* of a *class* whose definition specifies the private (internal) workings of these objects, as well as their public interface. Thus, in OOP, a class would be defined for each different type of object required. A class becomes a blueprint for building a particular kind of object. A class definition and appropriate *initial values* are used to create an instance (object) of the class. This operation is known as *object instantiation.*

The OOP technology also calls for easy ways to construct objects on top of other objects. There are two principal methods, *composition* and *inheritance.* Composition allows existing objects to be used as components to build other objects. For instance, a calculator object may be composed of an arithmetic unit object and a user interface object. Inheritance is a major OOP feature that allows you to extend and modify existing classes without changing their code.

In C++ and Java, this is done through class extension. A *subclass* can *inherit* code from its *superclass* and also add its own data and methods. For example, a graphics window, a text window, and a terminal emulator window can all be extended from a basic window class. Also, a check, an invoice, and an application form can all be extended from a basic business form class. Inheritance allows the extraction of commonalities among similar or related objects. It also allows classes in OO software libraries to be used for many different or unforeseen purposes. Inheriting from one class is *single inheritance*, and from several classes is *multiple inheritance.*

OOP also makes program libraries more useful. While traditional program libraries increase reusability, they are cast in concrete and can only function and be used in predefined ways. OOP libraries, on the other hand, can be modified through inheritance and are much more flexible when applied.

10.11 Object-Oriented Design

Program implementation is usually preceded by a design phase. For OOP, we would use *object-oriented design* (OOD). With OOD, the solution of a given problem or task is broken into self-contained, interacting objects that correspond to actual or logical entities in the solution process.

In a sense, OOD is like setting up a company for a particular purpose. The company contains a number of autonomous divisions, departments, factories, centers, and so on. Some of these entities contain other entities, and all have well-defined external behaviors and internal organizations. A company can adapt to many different tasks and react to a changing marketplace because these entities can adapt to different patterns of interaction without major reorganization. A well-designed OO program would be robust and agile just like a company.

As computer programs advance recursively and expansively, we, the users, must also keep up. We need to get them on our computers and get them in our minds to use them effectively.

Exercises

10.1. What operations can you perform on an application installed on your computer?

10.2. Give at least five ways to invoke an application.

10.3. What support does an application usually need from the OS?

10.4. What is an instruction set? How is machine language related to an instruction set?

10.5. What is an assembler? What is a mnemonic?

10.6. What is a compiler? an interpreter?

10.7. What makes a programming language high-level?

10.8. Name and explain at least five desirable qualities for a program.

10.9. What is modularity? Why is it good practice for programming?

10.10. What is OOP? What languages support OOP?

10.11. What OOP construct supports definition of objects? Extending object definitions?

10.12. What is API? What is it for?

10.13. What are the distinct phases for program compilation?

10.14. Explain the meaning of "programs help programming" and give examples.

10.15. **Computize**: Name three applications, outside the standard ones that came with your computer, that are important for you or will make your life easier.

10.16. **Computize**: What problems do you think cooperating processes need to address to perfectly coordinate their activities?

10.17. **Computize**: What applications would you run at once to help you perform a task or achieve a goal? Describe why you need them running at the same time.

10.18. **Computize**: In what language can a compiler and an assembler be written? Please explain.

10.19. **Group discussion topic**: *The options and settings of your smartphone camera.*

10.20. **Group discussion topic**: *Producer–consumer relationships.*

10.21. **Group discussion topic**: *The apps I love.*

10.22. **Group discussion topic**: *I want to become a computer professional.*

10.23. **Group discussion topic**: *What is the definition and scope of "computational thinking"? Does it matter?*

10.24. **Group discussion topic**: *My take away from this book.*

Epilogue

The broad and sometimes in-depth overview of computing in this book has provided you with key information and understanding to take better advantage of the digital age in which we live.

At school, at work, and at home, your growing set of CT tools help you computize and achieve greater success. Let's all

Always think computing and computize thinking.

Here is a final take away: Any success or setback we witness or experience is always a lesson to improve ourselves.

CT: REPROGRAM YOUR BRAIN

Alter your own programming by internalizing lessons at every single opportunity.

By continually forming new routines, establishing better habits, and increasing efficiency, you can reinvent yourself, in big and small ways, to achieve "SELF 2.0," recursively.

Please feel free to visit the CT website (`computize.org`) and share your own CT insights or contact the author with your feedback. The fact that you are reading this says you are serious about CT. You'll be the one reaping the rewards!

You, the readers, will be the core of a growing movement to spread the CT message widely at every level of society.

Website and Online Examples

Website

This book has a website (the CT website) useful for instructors and students:

`http://computize.org`

The CT website provides hands-on interactive demos, CT headlines, full-color figures, reader feedback, information updates, and instructor materials, as well as other resources for the textbook.

To request access to protected materials on the CT website, use the access code **Ct2015Fall**.

Bibliography

[1] V. Barr and C. Stephenson. *Bringing Computational Thinking to K-12: What is Involved and What is the Role of the Computer Science Education Community?* Computer Science Teachers Association Magazine ACM Inroads, 2(1), pages 48–54 (2011)

[2] M. Guzdial. *Paving the way for computational thinking.* CACM 51(8):25–27, (2008).

[3] J. J. Lu and George H. L. Fletcher. *Thinking about computational thinking.* Proceedings of the 40th ACM technical symposium on computer science education, ACM, ISBN: 978-1-60558-183-5, pages 260–264 (2009).

[4] D. Moursund and D. Ricketts. *Computational Thinking.* Information Age Education (IAE), `iae-pedia.org/Computational_Thinking` (2011).

[5] S. Papert. *Mindstorms: Children, Computers, and Powerful Ideas.* Basic Books, 2nd edition, ISBN: 0465046746 (1993).

[6] P. Phillips. *Computational Thinking A Problem-solving Tool for Every Classroom.* CSTA (Computer Science Teachers Association) (2009).

[7] J. M. Wing. *Computational thinking.* CACM 49(3):33–35 (2006).

[8] Committee for the Workshops on Computational Thinking. *Report of a Workshop on The Scope and Nature of Computational Thinking.* National Research Council, ISBN: 0-309-14958-4, 114 pages (2010).

[9] Committee for the Workshops on Computational Thinking. *Report of a Workshop on The Scope and Nature of Computational Thinking.* The National Academies Press, Washington, DC, ISBN 0-309-21474-2 (2011).

[10] Kojo Nnamdi Show. *Thinking Like a Computer Scientist*, interview with Dr. J. M. Wing. `thekojonnamdishow.org` (Nov. 18 2008).

[11] Resources at Google. *Exploring Computational Thinking.* Online materials at `www.google.com/edu/resources`

Index